Axiomization of Passage from "Local" Structure to "Global" Object

Recent Titles in This Series

485 **Paul Feit,** Axiomization of passage from "local" structure to "global" object, 1993
484 **Takehiko Yamanouchi,** Duality for actions and coactions of measured groupoids on von Neumann algebras, 1993
483 **Patrick Fitzpatrick and Jacobo Pejsachowicz,** Orientation and the Leray-Schauder theory for fully nonlinear elliptic boundary value problems, 1993
482 **Robert Gordon,** G-categories, 1993
481 **Jorge Ize, Ivar Massabo, and Alfonso Vignoli,** Degree theory for equivariant maps, the general S^1-action, 1992
480 **L. Š. Grinblat,** On sets not belonging to algebras of subsets, 1992
479 **Percy Deift, Luen-Chau Li, and Carlos Tomei,** Loop groups, discrete versions of some classical integrable systems, and rank 2 extensions, 1992
478 **Henry C. Wente,** Constant mean curvature immersions of Enneper type, 1992
477 **George E. Andrews, Bruce C. Berndt, Lisa Jacobsen, and Robert L. Lamphere,** The continued fractions found in the unorganized portions of Ramanujan's notebooks, 1992
476 **Thomas C. Hales,** The subregular germ of orbital integrals, 1992
475 **Kazuaki Taira,** On the existence of Feller semigroups with boundary conditions, 1992
474 **Francisco González-Acuña and Wilbur C. Whitten,** Imbeddings of three-manifold groups, 1992
473 **Ian Anderson and Gerard Thompson,** The inverse problem of the calculus of variations for ordinary differential equations, 1992
472 **Stephen W. Semmes,** A generalization of riemann mappings and geometric structures on a space of domains in $\mathbf{C}^n$, 1992
471 **Michael L. Mihalik and Steven T. Tschantz,** Semistability of amalgamated products and HNN-extensions, 1992
470 **Daniel K. Nakano,** Projective modules over Lie algebras of Cartan type, 1992
469 **Dennis A. Hejhal,** Eigenvalues of the Laplacian for Hecke triangle groups, 1992
468 **Roger Kraft,** Intersections of thick Cantor sets, 1992
467 **Randolph James Schilling,** Neumann systems for the algebraic AKNS problem, 1992
466 **Shari A. Prevost,** Vertex algebras and integral bases for the enveloping algebras of affine Lie algebras, 1992
465 **Steven Zelditch,** Selberg trace formulae and equidistribution theorems for closed geodesics and Laplace eigenfunctions: finite area surfaces, 1992
464 **John Fay,** Kernel functions, analytic torsion, and moduli spaces, 1992
463 **Bruce Reznick,** Sums of even powers of real linear forms, 1992
462 **Toshiyuki Kobayashi,** Singular unitary representations and discrete series for indefinite Stiefel manifolds $U(p,q;\mathbb{F})/U(p-m,q;\mathbb{F})$, 1992
461 **Andrew Kustin and Bernd Ulrich,** A family of complexes associated to an almost alternating map, with application to residual intersections, 1992
460 **Victor Reiner,** Quotients of coxeter complexes and P-partitions, 1992
459 **Jonathan Arazy and Yaakov Friedman,** Contractive projections in C_p, 1992
458 **Charles A. Akemann and Joel Anderson,** Lyapunov theorems for operator algebras, 1991
457 **Norihiko Minami,** Multiplicative homology operations and transfer, 1991
456 **Michał Misiurewicz and Zbigniew Nitecki,** Combinatorial patterns for maps of the interval, 1991
455 **Mark G. Davidson, Thomas J. Enright and Ronald J. Stanke,** Differential operators and highest weight representations, 1991
454 **Donald A. Dawson and Edwin A. Perkins,** Historical processes, 1991

(*Continued in the back of this publication*)

Number 485

Axiomization of Passage from "Local" Structure to "Global" Object

Paul Feit

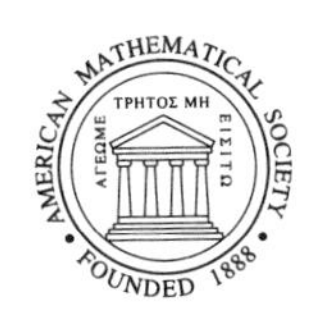

January 1993 • Volume 101 • Number 485 (end of volume) • ISSN 0065-9266

American Mathematical Society
Providence, Rhode Island

1991 *Mathematics Subject Classification.*
Primary 14A, 14K, 18A, 18B, 18D, 18F.

Library of Congress Cataloging-in-Publication Data

Feit, Paul, 1959–
Axiomization of passage from "local" structure to "global" object/Paul Feit.
p. cm. – (Memoirs of the American Mathematical Society; no. 485)
Includes bibliographical references.
ISBN 0-8218-2546-1
1. Geometry, Algebraic. 2. Categories (Mathematics) 3. Toposes. I. Title. II. Series.
QA3.A57 no. 485
[QA564]
510 s–dc20
[516.3′5] 92-33858
CIP

Memoirs of the American Mathematical Society

This journal is devoted entirely to research in pure and applied mathematics.

Subscription information. The 1993 subscription begins with Number 482 and consists of six mailings, each containing one or more numbers. Subscription prices for 1993 are $336 list, $269 institutional member. A late charge of 10% of the subscription price will be imposed on orders received from nonmembers after January 1 of the subscription year. Subscribers outside the United States and India must pay a postage surcharge of $25; subscribers in India must pay a postage surcharge of $43. Expedited delivery to destinations in North America $30; elsewhere $92. Each number may be ordered separately; *please specify number* when ordering an individual number. For prices and titles of recently released numbers, see the New Publications sections of the *Notices of the American Mathematical Society.*

Back number information. For back issues see the *AMS Catalog of Publications.*

Subscriptions and orders should be addressed to the American Mathematical Society, P. O. Box 1571, Annex Station, Providence, RI 02901-1571. *All orders must be accompanied by payment.* Other correspondence should be addressed to Box 6248, Providence, RI 02940-6248.

Memoirs of the American Mathematical Society is published bimonthly (each volume consisting usually of more than one number) by the American Mathematical Society at 201 Charles Street, Providence, RI 02904-2213. Second-class postage paid at Providence, Rhode Island. Postmaster: Send address changes to Memoirs, American Mathematical Society, P. O. Box 6248, Providence, RI 02940-6248.

Printed in the United States of America.
This volume was printed directly from author-prepared copy.
The paper used in this book is acid-free and falls within the guidelines
established to ensure permanence and durability. ♾

10 9 8 7 6 5 4 3 2 1 98 97 96 95 94 93

Table of Contents

Abstract

This paper offers a systematic approach to all mathematical theories with local/global behavior. To build objects with local and global aspects, one begins with a category $\mathfrak{C}$ of allowed local structures, and somehow derives a category $\mathfrak{C}^{gl}$ of things which are 'locally' in $\mathfrak{C}$. Some global objects, such as manifolds or schemes, can be represented as a sheaf of algebras on a topological base space; others, like algebraic spaces, are more technical. These theories share common structure—certain theorems on inverse limits, descent, and dependence on a special class of morphism (e.g., open embeddings) appear in all cases. Yet, classical proofs for universal properties proceed by case-by-case study. Separate examples require distinct arguments.

The present work places all local/global theories in a single, universal format. We define a local structure to be a category in which each object has a Grothendieck topology and to which a list of categorical axioms apply. (The formulation does not require models involving base spaces.) For a local structure $\mathfrak{C}$, we construct another local structure $\mathfrak{C}^*$ and a functor $* : \mathfrak{C} \longrightarrow \mathfrak{C}^*$ such that:

(1) For any classical choice of $\mathfrak{C}$, $\mathfrak{C}^*$ is the classical category of all 'locally $\mathfrak{C}$' objects.

(2) The universal propositions can be proved for $\mathfrak{C}^*$ in complete generality.

(3) The functor $* : \mathfrak{C} \longrightarrow \mathfrak{C}^*$ admits three distinct universal properties.

Keywords: Algebraic Geometry, Category Theory, Topos Theory

Introduction

Our purpose is to announce a foundational result concerning all mathematical notions of local/global behavior. The notions of 'local' and 'global' are common in any study with geometric aspects. Roughly, a global object is something characterized by the constraint that, locally, it restricts to a member of some specified category. The archetypal example is the C^∞-manifold, which is characterized by the property that local pieces of it are identified with open subsets of Euclidean space. The involved definition of another example, the scheme, merely formalizes the intuition that it be a thing derived by 'gluing' together commutative rings. Development of topos theory was partially motivated by a desire to carry the idea of local/global structure to more abstruse constructions.

The author will prove a universal statement in the following format: Given a category $\mathfrak{C}$, that one wishes to serve as the collection of all local objects for a theory, there is a list of axioms on $\mathfrak{C}$ under which one produces a category $\mathfrak{C}^*$ and a functor $^* : \mathfrak{C} \longrightarrow \mathfrak{C}^*$ such that:

(1.a) For any classical choice of $\mathfrak{C}$, $\mathfrak{C}^*$ is functorially equivalent to the classical category of all 'locally $\mathfrak{C}$' objects.

(1.b) Certain propositions, which are known to hold in each classical local/global theory, can be proved for $\mathfrak{C}^*$ in complete generality.

(1.c) The functor $^* : \mathfrak{C} \longrightarrow \mathfrak{C}^*$ is characterized by three distinct universal properties.

To illustrate the framework, let us discuss some examples.

The new format makes rigorous the pattern of development common to local/global theories. In a traditional approach, one starts with a category $\mathfrak{C}$, and a somehow builds from it a new category $\mathfrak{C}^{gl}$ of 'locally-$\mathfrak{C}$' objects. Essentially, an object of $\mathfrak{C}^{gl}$ is defined as anything which locally resembles a member of $\mathfrak{C}$. However, translation of the intuition into proper mathematics may be difficult, and different examples rely on

Received by editor April 5, 1991, and in revised form 10/1/91.

The work on this paper was partially supported by NSF Grant DMS 8601130.

different tricks. A crucial step for construction of $\mathfrak{C}^{gl}$ is formulation of a canonical functor $\Phi : \mathfrak{C} \longrightarrow \mathfrak{C}^{gl}$. Usually, Φ begins as a construction on each $\mathfrak{C}$-object, and then $\mathfrak{C}^{gl}$ is created as a coherent codomain for the things that Φ produces. In some imprecise way, all behavior of $\mathfrak{C}^{gl}$ is determined by $\mathfrak{C}$ and Φ. Examples to keep in mind are

Example A: $\mathfrak{C}^{gl}$ is the category of all C^∞-manifolds (of all dimensions). Here, $\mathfrak{C}$ is the category whose objects are pairs (U,n) where $n \in \mathbb{N}$ and $U \subseteq \mathbb{R}^n$ is open. Only C^∞-functions are regarded as $\mathfrak{C}$-morphism.

Example B: $\mathfrak{C}^{gl}$ is the category of schemes, and $\mathfrak{C}$ is the opposite category of commutative rings. The usual construction of Φ (and of $\mathfrak{C}^{gl}$) is indirect. First, a category $\mathcal{O}$ of locally ringed spaces is defined. Into $\mathcal{O}$ is a functor Φ_0 sending each $A \in \mathfrak{C}$ to its spectral sheaf. Since $\mathcal{O}$ is a category of sheaves (on topological base spaces), the term 'local' has rigorous meaning within it. $\mathfrak{C}^{gl}$ is realized, literally, as the subcategory of all locally-$\mathfrak{C}$ objects.

Example C: $\mathfrak{C}^{gl}$ is the category of rigid analytic spaces. Here, $\mathfrak{C}$ is an opposite category consisting of complete topological rings. As with the previous example, construction of Φ stems from the idea of a spectrum. Again, each member of $\mathfrak{C}$ is modeled as a base space with additional structure. However, to preserve analytic continuation, one cannot describe the structure as a sheaf on all open subsets.

Example D: See [D]. $\mathfrak{C}^{gl}$ is the category of Douady's espaces analytique banachique, essentially infinite dimensional complex-analytic manifolds. Here, $\mathfrak{C}$ is the category of pairs (U,E) where E is a complex Banach space and $U \subseteq E$ is open. The context may seem to be a slight variation on Example A. Yet, manifolds have models as sheaves on topological bases, which the espaces banachique lack. Douady observes that $\mathfrak{C}^{gl}$ must be built using the topos-theoretic version of sheaf. After establishing the key structural lemmas, he leaves miscellaneous verifications to the reader as 'asinitrottante'.

Example E: $\mathfrak{C}^{gl}$ is the category of algebraic spaces. Again $\mathfrak{C}$ is the opposite category of rings, but now it is assigned the *étale* topology. This situation is far outside the scope of sheaves over base spaces.

Intuitively, the relationship between $\mathfrak{C}$ and $\mathfrak{C}^{gl}$ is identical in all cases. The similarity is mathematical as well.

(2.a) In all cases, Φ preserves finite inverse limits.

(2.b) In all cases, for $A,B \in \mathfrak{C}$, Φ determines a bijection
$\mathrm{Mor}_{\mathfrak{C}}(A,B) \longrightarrow \mathrm{Mor}_{\mathfrak{C}^{gl}}(\Phi(A),\Phi(B))$.

(2.c) In most examples, there is a cotangent bundle construction. In fact, there are entire theories of 'coherent' bundles.

(2.d) In all cases, there are special morphisms, usually called *open embeddings* (or, in Example E, *étale* maps). These share certain, category-independent, properties. For example, consider the following commutative diagram in $\mathfrak{C}^{gl}$:

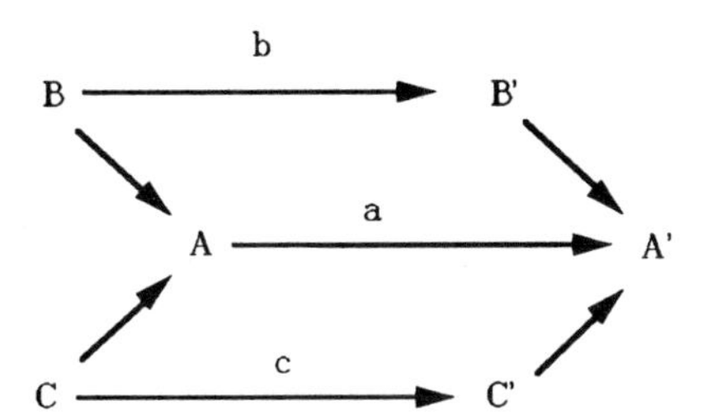

If a, b and c are open embeddings, and the corresponding products exist, then $b\times_a c : B\times_A C \longrightarrow B'\times_{A'} C'$ is an open embedding.

Classical methods do not recognize the underlying themes. Each example has a specific construction. Consequently, in each context, demonstration of each proposition (2.a,b,c,d) requires a distinct proof; that is, an argument which explains why certain structure in $\mathfrak{C}$ is preserved by the formalism particular to that example. Yet, these propositions merely state fundamental intuitions about the relationship between local and global structure.

The new approach places all the examples in one framework. Roughly, the universal theory consists of:

Step I: Let $\mathfrak{C}$ be a category in which every object has a Grothendieck topology. The first problem is to find conditions—phrased in terms of categorical operations, like pullback—on $\mathfrak{C}$ under which it becomes the basis of a local/global theory. We refer to such a category as a local structure.

The actual axioms are determined by needs of later proofs. Actually, the first step is the *conviction* that there *must exist* such a good list of axioms—that is, belief that both the category $\mathcal{C}^{gl}$ and the connecting functor Φ can be created purely from $\mathcal{C}$.

Step II: In each classical example, there is a procedure for 'building' global objects, usually referred to as a *cut-and-paste* or *descent* argument. The key to our theory is a formal, category-independent description of cut-and-paste.

Now define a global structure to be a local structure $\mathcal{D}$ with the added property that every cut-and-paste process with $\mathcal{D}$ objects yields another $\mathcal{D}$ object. For $\mathcal{C}$ a local structure, define a globalization for $\mathcal{C}$ to be a category $\mathcal{D}$ with a covariant functor $\Phi : \mathcal{C} \longrightarrow \mathcal{D}$ such that

- (i) $\mathcal{D}$ is a global structure,
- (ii) Φ is continuous (in some technical sense), and
- (iii) $\Phi : \mathcal{C} \longrightarrow \mathcal{D}$ is universal (ie., up to functorial equivalence) among functors which satisfy (i) and (ii).

The Main Problem is existence and characterization of globalizations.

Step III: Our results consist of a Main Theorem with useful corollaries. For $\mathcal{C}$ a local structure, we prove existence of a category $\mathcal{D}$ with a covariant functor $\Phi : \mathcal{C} \longrightarrow \mathcal{D}$ such that

- (i) $\mathcal{D}$ is a local structure,
- (ii) Φ is continuous,
- (iii) every cut-and-paste process in $\mathcal{C}$ yields an object in $\mathcal{D}$, and
- (iv) $\Phi : \mathcal{C} \longrightarrow \mathcal{D}$ is universal among functors which satisfy (i), (ii) and (iii).

We refer to such a universal functor as a *plus* functor, and denote it by $+ : \mathcal{C} \longrightarrow \mathcal{C}^{+}$. It is a stepping stone to globalization.

The following remarks are consequences of an explicit construction for plus functors.

- (i) A plus functor is fully faithful and preserves certain direct and inverse limits. It also admits secondary universal properties.
- (ii) The defining property of a plus functor $+ : \mathcal{C} \longrightarrow \mathcal{C}^{+}$ contains a

criterion under which a functor on $\mathfrak{C}$, such as a cotangent construction, has a canonical lift to $\mathfrak{C}^+$. In fact, it has several lifting properties.

(iii) For $\mathfrak{C}$ any local structure, the double iterate $\mathfrak{C} \longrightarrow \mathfrak{C}^{++}$ *must* be a globalization.

The last corollary implies existence of an abstract globalization for each classical Example. In addition, for each example, the theory of plus functors provides *trivial* proofs that traditional models of global objects have the universal property.

The above viewpoint has several advantages. All local structure of $\mathfrak{C}^{gl}$ (or $\mathfrak{C}^+$) is distilled to verification of conditions on $\mathfrak{C}$. In practice, $\mathfrak{C}$ is far simpler to work with than $\mathfrak{C}^{gl}$. The method trivializes issues on lifting functors. In some cases, it simplifies constructions involving base spaces. For example, consider the process by which each rigid analytic space is associated to a geometric base space. The first step is to assign a base space to each affinoid algebra; regard this as a functor Θ from the initial category $\mathfrak{C}$ to **Top**, the category of topological spaces. In a classical approach, one then characterizes each $A \in \mathfrak{C}$ as $\Theta(A)$ paired with additional structure. Demonstration that the category of such things is closed under cut-and-paste leads to the realization of rigid analytic spaces as bases with extra structure. Yet, to say that each member of $\mathfrak{C}^{gl}$ has a natural base space is just the claim that Θ has a canonical lift to a functor $\Theta : \mathfrak{C}^{gl} \longrightarrow \mathbf{Top}$. In the universal approach, the latter claim reduces to the simple remark that Θ is a continuous functor. The deeper issue, as to how much more data must be added to a base space before the entire analytic space can be recaptured, is irrelevant. In practice, an analytic space is treated as a collection of affinoid algebras linked together; the actual model, as base with structure, is used only to show that a universe of such objects is mathematically consistent.

The universal framework is an effort to simplify existing mathematical theories. Aside from the statement that construction is universal, the theory here has little to add directly to the established Examples A-E. Instead, we hope it can unify the multitude of frameworks used to build new objects by gluing charts.

As with much category theory, the essence of the work lies in the choice of axioms. Before abandoning the reader to pure formalism, we shall try to illustrate the two ideas which underlie everything.

Let $\mathcal{T}$ be a category. For us, a categorical topology for $\mathcal{T}$ is an assignment to each $\mathcal{T}$ object of a Grothendieck topology. Without going into details, we remark that definition of a categorical topology requires choice of a class Sub of $\mathcal{T}$-morphisms, which we call formal subsets. These morphisms generalize open embeddings and étale maps, as mentioned in (2.d).

Let $A \in \mathcal{T}$, and let $\mathcal{A} = \{\alpha_j : A_j \longrightarrow A\}_{j \in J}$ be a cone into A. Construct a graph $\mathbb{G} = \mathbb{G}(\mathcal{A})$ consisting of

(3) the objects A_j for $j \in J$,
choices of fibered products $A_j \times_A A_k$ for each $(j,k) \in J^2$,
all projections of the form $A_j \times_A A_k \longrightarrow A_j$ and $A_j \times_A A_k \longrightarrow A_k$ for $j,k \in J$.

There is a unique cone α from $\mathbb{G}$ to A which assigns α_j to each index j. Suppose $\mathcal{T}$ is assigned a topology such that if $\mathcal{A}$ is a cover (in the sense of Grothendieck), then the cone $\alpha : \mathbb{G} \longrightarrow A$ of $\mathcal{A}$ must be a colimit. We refer to such a topology as intrinsic. All motivations derive from intrinsic topologies.

Such graphs really do describe global objects. Consider the situation of Example B. Let $\mathbf{Ring}^0$ be the opposite category of rings, let **LR** be the category of locally ringed spaces, and let **Sch** be the category of schemes, realized inside **LR**. Our initial $\mathfrak{C}$ is $\mathbf{Ring}^0$, and we let $\mathcal{T} = \mathfrak{C}^{gl}$ be **Sch**. Suppose A is $\mathbb{P}^n$ for some $n \in \mathbb{N}$, and let the α_js be the n+1 canonical charts of A. Put $\mathbb{G}_0 = \mathbb{G}(\mathcal{A})$. All the fibered products are affine, so $\mathbb{G}_0$ becomes a graph of rings, inside the original $\mathfrak{C}$. All mapping and topological properties of $\mathbb{P}^n$ can be deduced from $\mathbb{G}_0$.

The analysis begins with the question:

(4) Suppose $\mathfrak{C}$ is a category with a Grothendieck topology and $\mathbb{G}$ is a graph of objects in $\mathfrak{C}$. Under what conditions (where each condition must be phrased in terms of $\mathfrak{C}$ and its topology) should $\mathbb{G}$ be regarded as the graph derived from a cover of a 'global' object which is 'locally $\mathfrak{C}$'?

Let $\mathfrak{G}$ be a graph in some category. For $\mathfrak{G}$ to be the canopy of a cover, the first, obvious, constraint is that its objects be indexed by $J \amalg J^2$, where J is some set that we will denote by $\underline{\Delta(\mathfrak{G})}$. The conjectured cover is indexed by J, and, for each index j, the j-th object of $\mathfrak{G}$ is the domain of the j-th member of the cover. But these hypothetical morphisms do not yet have a codomain; instead, $\mathfrak{G}$ offers information on their fibered products. The hard part of (4) is to augment these obvious points with axioms under which all the conjectured structure can be made formal.

In (4), it is *not* assumed that $\mathfrak{C}$ sits inside a large category (e.g., of sheaves) in which global objects exist. The problem is to recognize a cover of a global object *before* that object has been built! The above $\mathfrak{G}_0$ suggests the subtlety of (4). After all, $\mathfrak{G}_0$ has a colimit in the original $\mathfrak{C}$—namely $\mathbb{Z}$. Every colimit in $\mathfrak{C}$, which arises from a cover, remains a colimit in the larger category $\mathfrak{C}^{gl}$ which is under construction; why then, must the colimit of $\mathfrak{G}_0$ *not* be preserved, but instead replaced by a new global object, outside $\mathfrak{C}$? Moreover, $\mathfrak{G}_0$ not only determines the object $\mathbb{P}^n$ as a colimit, but also determines its topology. This suggests that the Grothendieck topology of $\mathfrak{C}^{gl}$ is *intrinsicly* determined by the topology of $\mathfrak{C}$. Yet, the classical definition of topology for a scheme is inherited from a larger category of spaces with a sheaf of rings.

Constructions via spaces with a sheaf, such as the spectral construction from $\mathbf{Ring}^0$ into **LR**, avoid (4). Instead, the initial $\mathfrak{C}$ is embedded into a category $\mathcal{D}$ in which each object carries a topological base. This $\mathcal{D}$ is chosen so that (a) its topology happens to be correct and (b) if $\mathfrak{G}$ is a graph of $\mathfrak{C}$-objects which represents a global object, then $\mathfrak{G}$ already has a colimit in $\mathcal{D}$. Roughly speaking, introduction of $\mathcal{D}$ resolves half of the issues of (4) by giving an explicit solution; the remaining details are handled by ad hoc lemmas.

Again, consider the usual approach to schemes. In many ways **LR** confuses the construction of **Sch**. Suppose, for example, B and C are schemes over another scheme A. The fibered product $B \times_A C$ in Sch is built by gluing together fibered products of affine charts. It bears no relation to the corresponding fibered product inside **LR**, *unless* B or C is restriction to a subset of A's base. For some reason, the spectral model is sufficient for modeling schemes, though it does not preserve fibered products. Mysteriously, to prove existence of $B \times_A C$ in general, it is very important that the two notions of product happen to agree when B is restriction of A.

In the present work, we offer a solution to (4). A graph of the appropriate type is called a(n abstract) canopy. To make sense of the question, we must adjust slightly the usual notion of a Grothendieck topology.

The second heuristic is more technical. In what follows, let $\mathfrak{C}$ be a topologized category, and let $\mathfrak{C}^+$ be the class of all abstract canopies.

Imagine that $\mathfrak{C}$ lies inside a global structure $\mathfrak{C}^{gl}$. For each $\mathbb{G} \in \mathfrak{C}^+$, there is a colimit $\theta : \mathbb{G} \longrightarrow X$ *in* $\mathfrak{C}^{gl}$. Identifying $\mathbb{G}$ with its limit gives a map from $\mathfrak{C}^+$ into $\mathfrak{C}^{gl}$. There is a class-theoretic function $+ : \mathfrak{C} \longrightarrow \mathfrak{C}^+$, namely map $A \in \mathfrak{C}$ to the canopy A^+ of the cover consisting of the lone morphism $1_A : A \longrightarrow A$; obviously, this assignement corresponds to embedding $\mathfrak{C} \longrightarrow \mathfrak{C}^{gl}$. Turning the remark on its head offers an approach for building $\mathfrak{C}^{gl}$: Can one introduce morphisms between members of $\mathfrak{C}^+$, which makes $\mathfrak{C}^+$ a category, and a topology for $\mathfrak{C}^+$ such that $\mathfrak{C}^+$ becomes a category filled with locally-$\mathfrak{C}$ objects? The answer to this question is the Main Theorem.

The program is

(5.a) assign to $\mathfrak{C}^+$ a notion of morphism,

(5.b) assign to $\mathfrak{C}^+$ a topology,

(5.c) show that $+$ is a functorial embedding with a universal property.

In fact, the vague objective (5.c) controls how we make decisions (5.a,b).

Consider possible topologies for some $\mathbb{G} \in \mathfrak{C}^+$. Recall that $\mathbb{G}$ is interpreted both as a global object *and* as the canopy of some cover for that object. In particular, for each $j \in \Lambda(\mathbb{G})$, there is a canonical formal ($\mathfrak{C}^+$-)subset $\iota_j : \mathbb{G}[j]^+ \longrightarrow \mathbb{G}$. Any morphism $B^+ \longrightarrow \mathbb{G}$ which composes an ι_j with a formal subset of $\mathfrak{C}$ (under $+$) must be a formal subset in $\mathfrak{C}^+$; call such morphisms the affine subsets. Our interpretation requires that $\{\iota_j\}_{j \in \Lambda(\mathbb{G})}$ be a cover for $\mathbb{G}$ in $\mathfrak{C}^+$. That this cone be a cover implies that certain other cones, consisting of affine subsets, must also be covers; call these the affine covers. In summary, we begin with the knowledge that certain $\mathfrak{C}^+$-morphisms *must* be formal subsets and that certain cones *must* be covers.

Suppose we are deciding whether a $\mathfrak{C}^+$-morphism $f : \mathbb{G} \longrightarrow \mathbb{H}$ should be a formal subset in the new topology. It is reasonable to require that there be an affine cover S of $\mathbb{G}$ such that $f \circ s$ is an affine subset of $\mathbb{H}$ for each $s \in S$. This is an easy criterion for recognition of a morphism which is *locally* a formal subset. Alas, there are many classical examples of local embeddings which fail to embed globally. Here then is a new problem (or, to students of topos theory, a variant of an old one): if we work with just the minimal axioms for a Grothendieck topology, then the topology on $\mathfrak{C}^+$—that is, which morphisms are open embeddings, and which families cover—will not be uniquely determined by properties such as (2.a,b,c,d).

The issue is settled by adding requirements for a topology of a local structure. A single morphism which covers its codomain is called a covering morphism. In Examples A-D, only isomorphisms are covering morphisms. We postulate that a morphism $b : X \longrightarrow Y$ must be a formal subset provided (a) it is, locally, a formal subset, and (b) each projection $X \times_Y X \longrightarrow X$ is a covering morphism. (For technical reasons, the precise condition (b) is formulated in a slightly stronger form.) In Examples A-D, condition (b) becomes the well-known fact that a local embedding is a true embedding if and only if it is injective. For general $\mathfrak{C}$, the heuristic analysis suggests that the class of $\mathfrak{C}^+$-morphisms which satisfy (a) is completely characterized in terms of + functor. The criterion reduces the problem of defining formal subsets for $\mathfrak{C}^+$ to the ability to lift the notion of covering morphism from $\mathfrak{C}$ in a canonical manner.

The above criterion enhances the role for Cvm, the class of covering morphisms. To make lifting possible, we must further formulate 'rigidity' axioms upon Cvm. Among these is the notion that $\mathfrak{C}$ be closed under Cvm; this formality demands that for $X \in \mathfrak{C}$, any $\mathfrak{C}^+$-morphism $\mathfrak{G} \longrightarrow X^+$ which covers X is $\mathfrak{C}^+/X^+$-isomorphic to a covering morphism of $\mathfrak{C}$, under +.

Part I formulates precisely the majority of our axioms. Part II introduces canopies as graphs meeting certian mapping and topological constraints. An intermediary category $\mathfrak{C}^p$ is constructed, and a Grothendieck topology is assigned to it. The object class of $\mathfrak{C}^p$ is the class of all canopies of $\mathfrak{C}$ objects; however, the notion of $\mathfrak{C}^p$-morphism is too restrictive. The correct definition of morphism between two canopies requires many technical verifications, and $\mathfrak{C}^p$ offers an environment in which efficiently to develop lemmas. Of especial interest is the notion of pullback in $\mathfrak{C}^p$. Part III focuses on a different aspect of $\mathfrak{C}^p$. Emphasis is on morphisms (called reductions) which are not $\mathfrak{C}^p$-isomorphisms, but which should become isomorphisms as the category enlarges to $\mathfrak{C}^+$. Meaning is given to the idea of a category closed under a universe of morphisms, although little is done with this until Part V.

Part IV introduces a process which takes a category $\mathcal{D}$ with a non-intrinsic topology and adds morphisms to it to make a category $\mathcal{D}^s$ with a nearly intrinsic topology. The process is called 'smoothing'. Of interest to us is the choice $\mathcal{D} = \mathfrak{C}^{\mathrm{p}}$. In this case, $\mathcal{D}^{\mathrm{s}}$ is the desired, intrinsic $\mathfrak{C}^{+}$. Fibered products in $\mathcal{D}^{\mathrm{s}}$ are characterized, as well as universal properties with respect to special types of functors.

Part V finally combines the miscellaneous postulates into definitions for local structure and global structure. Theorem 14.4 summarizes our results on the plus functor and on globalization. It is proved in Section 16. Section 15 interpolates the method for lifting a universe of layered morphisms from $\mathfrak{C}$ to $\mathfrak{C}^{+}$.

Index of Terms

Our approach juggles many kinds of constraints. Below is an index for terms formulated in the process. Each is assigned the number of equation, proposition or definition preceeding its introduction.

PART I TERMINOLOGY

§1 Standard Notations

§1.A Basics

Let $\mathbb{N}$ denote the set of natural numbers. For $n \in \mathbb{N}$, put

(1.1) $$\mathbb{N}(n) = \{ m \in \mathbb{N} : m \leq n \}.$$

Indexed families of objects are treated as functions. Typically, an indexed family is denoted by a single symbol θ, its indexing set is referred to as $\mathrm{dom}(\theta)$ or the domain of θ, the *set* of objects in θ by $\mathrm{Im}(\theta)$, and the j-th object by $\theta(j)$ for each $j \in \mathrm{dom}(\theta)$. A set S is freely regarded as the identity function on S. An indexed family θ is called non-empty if $\mathrm{dom}(\theta) \neq \emptyset$.

§1.B Axioms for Categories and Graphs

We adopt the Gödel-Bernays axioms for set theory (see [Göl]). In particular, the concept of sets extends to entities called 'classes'. Classes are treated like sets, but play the role of collections too large to be sets. There is a class of all sets.

A <u>metagraph</u> consists of the following data:

(1.2.a) a class $\mathcal{O}$ of <u>objects</u>,

(1.2.b) a class $\mathcal{M}$ of <u>morphisms</u>,

(1.2.c) two (class-theoretic) functions $\mathcal{M} \longrightarrow \mathcal{O}$ called the <u>domain</u> and <u>codomain</u> maps.

Suppose $\mathcal{C}$ is a metagraph. Denote its object class by $\mathcal{C}$ or $\mathrm{Obj}(\mathcal{C})$, its morphism class by $\mathrm{Mor}(\mathcal{C})$, and its domain and codomain functions by dom and cod respectively. For $X, Y \in \mathcal{C}$, put

(1.3) $$\mathrm{Mor}_{\mathcal{C}}(X,Y) = \{ f \in \mathrm{Mor}(\mathcal{C}) : \mathrm{dom}\, f = X \text{ and } \mathrm{cod}\, f = Y \},$$

(this is a class, though not necessarily a set). A <u>homomorphism</u> from a metagraph $\mathcal{C}$ to another $\mathcal{D}$ is a pair of functions $\mathrm{FO} : \mathrm{Obj}(\mathcal{C}) \longrightarrow \mathrm{Obj}(\mathcal{D})$ and $\mathrm{FM} : \mathrm{Mor}(\mathcal{C}) \longrightarrow \mathrm{Mor}(\mathcal{D})$ such that $(\mathrm{dom}_{\mathcal{D}}) \circ \mathrm{FM} = \mathrm{F\dot{U}} \circ (\mathrm{dom}_{\mathcal{C}})$ and $(\mathrm{cod}_{\mathcal{D}}) \circ \mathrm{FM} = \mathrm{FU} \circ (\mathrm{cod}_{\mathcal{C}})$; usually, one symbol is used to denote both maps.

Two types of metagraphs are of present interest. A <u>metacategory</u> is a metagraph with identity and composition assignments that satisfy the usual identites. A <u>category</u> is then a metacategory $\mathcal{C}$ in which

(1.4.a) both $\mathrm{Obj}(\mathcal{C})$ and $\mathrm{Mor}(\mathcal{C})$ are subclasses of the class of all sets, and

(1.4.b) for $X,Y \in \mathfrak{C}$, $\mathrm{Mor}_{\mathfrak{C}}(X,Y)$ is a set.

Covariant functors may be regarded as a special type of metagraph homomorphism. Throughout the paper, 'functor' means 'covariant functor'.

A metagraph S for which Obj(S) and Mor(S) are sets is called a graph. If S is a graph and $\mathfrak{C}$ is a category, then a metagraph homomorphism $S \longrightarrow \mathfrak{C}$ is called a graph of $\mathfrak{C}$-objects of type S.

To define a graph, we often give Obj(S) explicitly and express Mor(S) as a set of triples of the form (X,f,Y) where $X,Y \in \mathrm{Obj}(S)$ and f is a formal symbol. Our conventions are to freely denote (X,f,Y) by 'f', and to assign

(1.5) $\mathrm{dom}(X,f,Y) = X$ and $\mathrm{cod}(X,f,Y) = Y$.

When defining a category $\mathfrak{C}$, we give $\mathrm{Obj}(\mathfrak{C})$ explicitly, but generally define $\mathrm{Mor}_{\mathfrak{C}}(X,Y)$ for each $X,Y \in \mathfrak{C}$ rather than organize morphisms into the class $\mathrm{Mor}(\mathfrak{C})$.

Graphs and functors share a common notion of equivalence. Suppose $\mathcal{G}$ is a metagraph and $\mathfrak{C}$ is a category. For F and G metagraph homomorphisms $\mathcal{G} \longrightarrow \mathfrak{C}$, a transformation from F to G is a (class-theoretic) function which to each $x \in \mathrm{Obj}(\mathcal{G})$ assigns a $\mathfrak{C}$-morphism $\tau(x) : F(x) \longrightarrow G(x)$ such that

(1.6) $\forall f \in \mathrm{Mor}(\mathcal{G}), \quad \tau(\mathrm{cod}\, f) \circ F(f) = G(f) \circ \tau(\mathrm{dom}\, f)$.

Two such graphs are called equivalent if there is such a transformation τ which assumes $\mathfrak{C}$-isomorphisms. If both $\mathcal{G}$ and $\mathcal{D}$ are categories, we say a functor $F : \mathfrak{C} \longrightarrow \mathcal{D}$ is invertible if there is a functor $G : \mathcal{D} \longrightarrow \mathfrak{C}$ such that $F \circ G$ and $G \circ F$ are each equivalent to the appropriate identity functor.

§1.C Conventions in a Category

Let $\mathfrak{C}$ be a category.

A $\mathfrak{C}$-morphism f is called monomorphic, or a monomorphism if for each $X \in \mathfrak{C}$, the function

(1.7) $g \longmapsto f \circ g$ from $\mathrm{Mor}_{\mathfrak{C}}(X, \mathrm{dom}\, f) \longrightarrow \mathrm{Mor}_{\mathfrak{C}}(X, \mathrm{cod}\, f)$

is injective. A $\mathfrak{C}$-morphism is called epimorphic, or an epimorphism, if it is monomorphic with respect to the opposite of $\mathfrak{C}$.

Let $X \in \mathfrak{C}$. The category of pairs (B,b) where $B \in \mathfrak{C}$ and $b \in \mathrm{Mor}_{\mathfrak{C}}(B,X)$ is called the slice category over X, and is denoted by $\mathfrak{C}/X$. We frequently identify a $\mathfrak{C}$-morphism f into X with the $(\mathrm{dom}\, f, f) \in \mathfrak{C}/X$. By a cone into X, we mean any (indexed) family in $\mathfrak{C}/X$. The empty set is a cone into every object; however, when Θ is a non-empty cone into X, then we may set $X = \operatorname{cod}\theta$ without ambiguity.

§1.D Colimits, Inverse Limits and Pullbacks

Let $\mathfrak{C}$ be a category, let S be a graph, and let **F** be a graph of $\mathfrak{C}$-objects of type S.

By a cone from F, we mean a pair $(A;\theta)$ (also denoted by $\theta : F \longrightarrow A$) where $A \in \mathfrak{C}$ and θ is a function which to each $s \in S$ assigns $\theta(s) \in \operatorname{Mor}_{\mathfrak{C}}(F(s),A)$ such that for $g \in \operatorname{Mor}(S)$,

(1.8) $\theta(\operatorname{cod} g) \circ F(g) = \theta(\operatorname{dom} g)$.

If $\theta : F \longrightarrow A$ and $\varphi : F \longrightarrow B$ are cones and $h \in \operatorname{Mor}_{\mathfrak{C}}(A,B)$, then we say h factors φ through θ (or that h is a factoring) if $\varphi(s) = h \circ \theta(s)$ for each $s \in S$. A cone is a direct limit or colimit if every other cone of F has a unique factoring through it.

Inverse limits are the concept dual to colimits. We adopt similar terminology, except that a collection of morphisms from a $\mathfrak{C}$-object B to members of **F** which satisfies the condition dual to (1.8) is called a source for **F**.

Inverse limits figure prominantly in our discussion, so we introduce some shorthand. An inverse limit $(A;\theta)$ is identified with the object A, while each morphism of θ is denoted by 'π' with a suitable subscript or superscript. Suppose S is a graph which has a subset in $T \subseteq \operatorname{Obj}(S)$ such that

(1.9.a) for each $s \in S$, there is $g \in \operatorname{Mor}(S)$ so $s = \operatorname{cod} g$ and $\operatorname{dom} g \in T$,

(1.9.b) T is identified with $\mathbb{N}(n)$ for some $n \in \mathbb{N}$.

For binary products, $T = S = \{1,2\}$. A source (A,θ) for a graph of objects **F** of type S is determined by the restriction of θ to $T \approx \mathbb{N}(n)$; for this reason, θ is represented by the sequence $\theta(1),\ldots,\theta(n)$. If **F** is of type S, $(A;\theta)$ is a source of **F** and $(B;\varphi)$ is an inverse limit of **F**, then the unique factoring of θ through φ is denoted by $\theta(1) \wedge \cdots \wedge \theta(n)$.

Suppose $A \in \mathfrak{C}$ and $(B,b),(C,c) \in \mathfrak{C}/A$. The product of (B,b) and (C,c) in $\mathfrak{C}/A$ is called a fibered product over A, and is denoted by $(B,b) \times_A (C,c)$ or $B \times_A C$. This product is also called the pullback of (C,c) along b; in this case, it is denoted by $b^{-1}(C,c)$, its projection to B is denoted $b^{-1}c$, and its projection to C is expressed by $b_|$ or $b_{|b^{-1}C}$.

We shall repeatedly invoke the elementary

Proposition 1.10: Let $\mathfrak{C}$ be a category. Suppose we have the following commutative diagram of objects and morphisms from $\mathfrak{C}$:

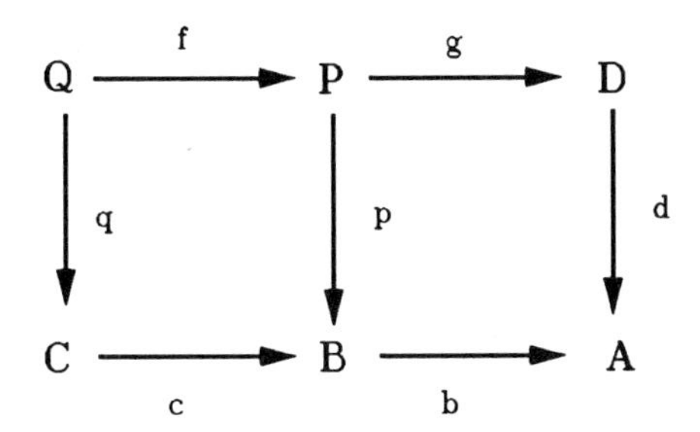

(1.11)

Suppose $(P;p,g)$ is a fibered product $(B,b)\times_A(D,d)$. Then $(Q;q,f)$ is a fibered product for $(C,c)\times_B(P,p)$ if and only if $(Q;q,g\circ f)$ is a fibered product for $(C,b\circ c)\times_A(D,d)$.

Proof: Trivial. □

Suppose

(1.12) $\quad A,A' \in \mathfrak{C},\quad (B,b),(C,c) \in \mathfrak{C}/A,\quad (B',b'),(C',c') \in \mathfrak{C}/A,$

$\quad h \in \mathrm{Mor}_{\mathfrak{C}}(A,A'),\quad f \in \mathrm{Mor}_{\mathfrak{C}}(B,B')\quad$ and $\quad g \in \mathrm{Mor}_{\mathfrak{C}}(C,C')$

such that

(1.13) $\quad b'\circ f = h\circ b\quad$ and $\quad c'\circ g = h\circ c.$

Then

(1.14) $\quad (f\circ\pi_B)\wedge(g\circ\pi_C) \in \mathrm{Mor}_{\mathfrak{C}/A}(B\times_A C, B'\times_{A'} C')$

is represented as $f\times_h g$.

Corollary 1.15: Suppose $A,B,M,Y,Z \in \mathfrak{C}$ and $a: A \longrightarrow B$, $b: B \longrightarrow M$, $z: Z \longrightarrow Y$ and $y: Y \longrightarrow M$ are $\mathfrak{C}$-morphisms. Suppose all of the indicated products in the following diagram exist:

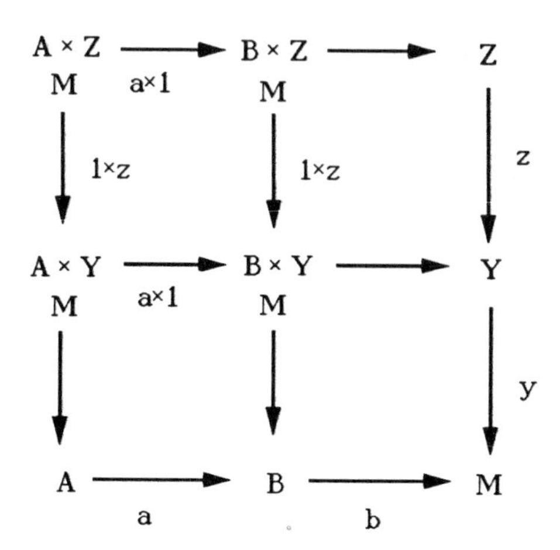

(1.16)

Then

(1.17.a) $(A\times_M Z; 1_A\times z, \pi_Z)$ is a fibered product $(A\times_M Y, \pi_Y)\times_Y (Z,z)$,

(1.17.b) $(A\times_M Z; \pi_A, a\times 1_Z)$ is a fibered product $(A,a)\times_B (B\times_M Z, \pi_B)$,

(1.17.c) $(A\times_M Z; 1_A\times z, a\times 1_Z)$ is a fibered product $(A\times_M Y, a\times 1_Y)\times_{B\times_M Y} (B\times_M Z, 1_B\times z)$.

Proof: Follows from repeated use of Proposition 1.10. □

Now suppose $\mathfrak{C}$ is a category, S is a non-empty graph, G is a graph of $\mathfrak{C}$-objects of type S, $\Theta : G \longrightarrow A$ is a cone and $b : B \longrightarrow A$ is a $\mathfrak{C}$-morphism. We freely identify G with the graph G' of $\mathfrak{C}/A$ objects by identifying G(s) with $(G(s),\Theta(s))$, when the meaning is clear. Suppose H is another graph of $\mathfrak{C}$ objects of type S, $\varphi : H \longrightarrow B$ is a cone and $\tau : G \longrightarrow H$ is a transformation of graphs such that

(1.18) for $s \in S$, $(H(s);\varphi(s),\tau(s))$ is a pullback $b^{-1}\Theta(s)$.

In this case we call H a (choice of) pullback of G along b, φ the pullback of Θ along b, and τ the canonical projection $H \longrightarrow G$. Such H exists provided that each of the pullbacks of (1.18) exists, and H is uniquely determined in the obvious sense.

Proposition 1.19: Suppose $\mathfrak{C}$ is a category, $A,B \in \mathfrak{C}$, $b \in \mathrm{Mor}_{\mathfrak{C}}(B,A)$ and S is a graph. Let G' and H' be graphs of type S consisting of $\mathfrak{C}/A$ and $\mathfrak{C}/B$ objects, respectively and assume H' is a pullback of G' along b.

Suppose $\alpha : A' \longrightarrow G'$ is a source in $\mathcal{C}/A$. Suppose B' is a fibered product $(B,b)\times_A A'$ along b; define a source $\beta : B' \longrightarrow H'$ by $s \mapsto 1_B \times \alpha(s)$. We call β the pullback of α (along b). If α is an inverse limit in $\mathcal{C}/A$, then β is an inverse limit in $\mathcal{C}/B$.

Proof: Trivial □

§2 Grothendieck Topologies

Grothendieck's axiomization of topology is the basis of the present work. However, our theory emphasizes unusual aspects of his formulation, so we develop a terminology with non-standard definitions.

§2.A Universes of Morphisms and Topologies

Topology leads to special 'types' of morphisms. An embedding, or covering map, or étale homomorphisms, has traditionally been regarded as a morphism with properties outside the language of category theory. But, it appears impossible to axiomatize cut-and-paste arguments without the ability to determine whether or not a given morphism is an open embedding. Our theory begins with the concept of a previously supplied family of 'good' morphisms.

Let Sub be a subclass of $\mathrm{Mor}(\mathfrak{C})$. Suppose that

(2.1.a) Sub contains all $\mathfrak{C}$-isomorphisms,

(2.1.b) the composition of any two members of Sub is again in Sub,,

(2.1.c) if $A \in \mathfrak{C}$ and $(B,b),(C,c) \in \mathfrak{C}/A$ so that $b \in \mathrm{Sub}$, then there exists a fibered product for $(C,c)\times_A(B,b)$; moreover, for $(P;\pi_C,\pi_B)$ such a fibered product, $\pi_C \in \mathrm{Sub}$.

We say that Sub is a <u>universe of subsets</u> for $\mathfrak{C}$. A member of Sub is called a <u>Sub-morphism</u> or a <u>formal subset</u>. For $A,B \in \mathfrak{C}$, the class $\mathrm{Mor}_{\mathfrak{C}}(A,B) \cap \mathrm{Sub}$ is denoted by $\mathrm{Sub}(A,B)$.

A $\mathfrak{C}$-morphism $d: D \longrightarrow A$ is called a <u>pullback base</u> (of $\mathfrak{C}$) if each member of $\mathfrak{C}/A$ has a pullback along it. Proposition 1.10 implies

<u>Proposition 2.2:</u> The class of pullback bases of $\mathfrak{C}$ is a universe of subsets.

Let Sub be a universe of subsets.

For an object $A \in \mathfrak{C}$, a subset $J \subseteq \mathfrak{C}/A$ is called a <u>choice of representatives</u> (for Sub over A) if

(2.3.a) for each $(B,b) \in J$, $b \in \mathrm{Sub}(B,A)$,

(2.3.b) for $C \in \mathfrak{C}$ and $c \in \mathrm{Sub}(C,A)$, (C,c) is $\mathfrak{C}/A$-isomorphic to a member of J.

We say that Sub satisfies the <u>smallness condition</u> if each object admits a choice of subsets.

Sub is called a <u>universe of embeddings</u> if

(2.4) for $c \in \mathrm{Mor}(\mathfrak{C})$ and $b \in \mathrm{Sub}$ so $\mathrm{dom}\, b = \mathrm{cod}\, c$, if $b \circ c \in \mathrm{Sub}$ then $c \in \mathrm{Sub}$.

Classical examples of embeddings meet this axiom. The word 'embedding' is usually reserved for monomorphisms. In fact, the monomorphic property implies condition (2.4). Suppose $c: C \longrightarrow B$ and $b: B \longrightarrow A$ are $\mathfrak{C}$-morphisms so that b is monomorphic and $b \circ c$ is a formal subset. Then c is a pullback of $b \circ c$ along b, and, therefore, is a formal subset!

The difference between universes of subsets and those of embeddings is illustrated by

Proposition 2.5: Let $\mathfrak{C}$ be a category, and let Sub be a universe of subsets for $\mathfrak{C}$.

(A) Let $A \in \mathfrak{C}$, $(B,b),(B',b'),(C,c),(C',c) \in \mathfrak{C}/A$, $f \in \mathrm{Sub}(B,B')$ and $g \in \mathrm{Sub}(C,C')$ such that $b' \circ f = b$ and $c' \circ g = c$. Suppose

(2.6) either b' or c' is a pullback base.

Then $f \times g : B \times_A C \longrightarrow B' \times_A C'$ is a Sub-morphism.

(B) Suppose Sub is a universe of embeddings. Let $A \in \mathfrak{C}$, $(B,b),(C,c) \in \mathfrak{C}/A$ and $a \in \mathrm{Sub}(A,A')$. If b or c is a pullback base, then $1_B \times_a 1_C : B \times_A C \longrightarrow (B, a' \circ b) \times_{A'} (C, a' \circ c)$ is a Sub-morphism.

Proof: Suppose $A \in \mathfrak{C}$, $(B',b'),(C,c) \in \mathfrak{C}/A$ and $f \in \mathrm{Sub}(B,B')$. Put $b = b' \circ f$. Proposition 1.10 and condition (2.1.c) assure that $f \times 1_C : B \times_A C \longrightarrow B' \times_A C$ is a formal subset, provided that both fibered products exist. The roles of B and C may be interchanged. Part (A) follows.

To prove (B), we need another diagram. Assume the hypothesis in which c is a pullback base.

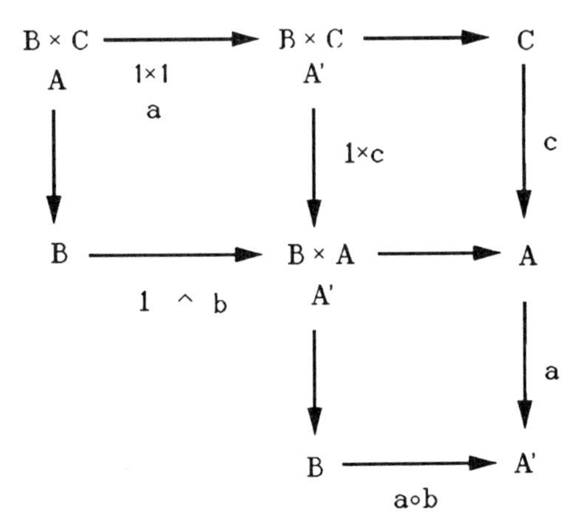

(2.7) $(B \times_{A'} C; 1_B \times c, \pi_C)$ is a fibered product $(B \times_{A'} A, \pi_A) \times_A C$,

$(B\times_A C;\pi_B,1_B\times_a 1_C)$ is a fibered product $(B,1\wedge b)\times_{(B\times_{A'}A)}(B\times_{A'}C,1_B\times c)$.

We are reduced to showing that $1\wedge b$ is an embedding. Now π_B on $B\times_{A'}A$ is an embedding because $a\in \mathrm{Sub}(A,A')$, and $\pi_B\circ(1\wedge b)=1_B$ is also an embedding. By (2.4), $1\wedge b$ must be an embedding. □

Let Θ be an indexed cone into $A\in\mathfrak{C}$. A subdivison function of Θ is a map which to each $j\in\mathrm{dom}(\Theta)$ assigns a cone φ_j into $\mathrm{dom}\,\Theta(j)$. Suppose φ is such a function. Put

(2.8) $\Lambda=\{(j,r)\ :\ j\in\mathrm{dom}(\Theta),\ r\in\mathrm{dom}(\varphi_j)\}$.

The function $\tau:(j,r)\mapsto\Theta(j)\circ\varphi_j(r)$ is called the subdivison of Θ through φ. We also refer to $\mathrm{Im}(\tau)$ as the subdivision.

Suppose Cov is a class of subsets of Sub such that

(2.9.a) each $S\in\mathrm{Cov}$ is a non-empty cone,

(2.9.b) if $S\in\mathrm{Cov}$ and T is a subset of Sub such that T is cone and $S\subseteq T$, then $T\in\mathrm{Cov}$,

(2.9.c) if b is a $\mathfrak{C}$-isomorphism, then $\{b\}\in\mathrm{Cov}$,

(2.9.d) if $S\in\mathrm{Cov}$ and Θ is a function which to each $s\in S$ assigns $\Theta(s)\in\mathrm{Cov}$ so that $\mathrm{cod}(\Theta(s))=\mathrm{dom}(s)$, then the subdivison of S through Θ is in Cov,

(2.9.e) if $S\in\mathrm{Cov}$, $B=\mathrm{cod}(S)$, $b:A\longrightarrow B$ is a $\mathfrak{C}$-morphism and Θ is a function which to each $s\in S$ assigns a pullback $\Theta(s)=(b^{-1}s;\pi_A,\pi_s)$, then $\{(b^{-1}s,\pi_A)\ :\ s\in S\}\in\mathrm{Cov}$,

(2.9.f) if $f\in\mathrm{Sub}$ and π_1 and π_2 are projections of $f\times_{\mathrm{cod}\,f}f$ to $\mathrm{dom}\,f$, then $\{\pi_1\},\{\pi_2\}\in\mathrm{Cov}$.

We call Cov a Grothendieck topology for Sub, and refer to $(\mathfrak{C},\mathrm{Sub},\mathrm{Cov})$ as a topologized category. A $\mathfrak{C}$-morphism b for which $\{b\}\in\mathrm{Cov}$ is called a covering morphism. An indexed family Θ of objects is called a(n indexed) cover if $\mathrm{Im}(\Theta)\in\mathrm{Cov}$; we also say that Θ covers $\mathrm{cod}\Theta$, or that Θ is a(n indexed) cover of this object.

There is another form of (2.9.f) which is convenient:

Proposition 2.10: Let $(\mathfrak{C},\mathrm{Sub},\mathrm{Cov})$ be a topologized category. Suppose $b:B\longrightarrow A$ is a formal subset and $s\in\mathrm{Mor}_{\mathfrak{C}}(A,B)$ such that $b\circ s$ is an isomorphism. Then b is a covering map.

Proof: Without loss of generality, assume $b\circ s=1_A$. Let $(C;\pi_1,\pi_2)$ be a fibered product $b\times_A b$.; then both π_1 and π_2 are covering morphisms. Proposition 1.10 allows us to interpret $b:B\longrightarrow A$ as projection $(C,\pi_2)\times_B(A,s)\longrightarrow A$. As π_2 is a covering morphism, b must cover as well. □

Conversely, suppose $b: B \longrightarrow A$ is in Sub, and let δ be the diagonal map $1_B \wedge 1_B: B \longrightarrow B \times_A B$. Composition of δ with either projection is 1_B, and hence the projections meet the hypothesis of Proposition 2.10.

Suppose Cov is a topology of Sub. For $A \in \mathcal{C}$, a subset $J \subseteq \mathcal{C}/A$ is called a <u>choice of representatives with topology</u> (for (Sub,Cov) over A) if

(2.11.a) J is a choice subsets over A for Sub,

(2.11.b) the class T, of all subsets $S \in$ Cov which are subsets of J, is a set.

Cov satisfies the <u>smallness condition</u> if each $A \in \mathcal{C}$ admits a choice of representatives with topology. A class-theoretic function which assigns to each $A \in \mathcal{C}$ a choice of representatives with topology over A is called a <u>categorical choice of representatives with topology</u> for $\mathcal{C}$.

For the rest of this section, assume ($\mathcal{C}$,Sub,Cov) is a topologized category.

The universe of subsets associated to $\mathcal{C}$ and its topology are denoted respectively by Sub (or $\mathrm{Sub}_{\mathcal{C}}$) and Cov (or $\mathrm{Cov}_{\mathcal{C}}$). The class of covering morphisms is another universe of subsets for $\mathcal{C}$, which is denoted by Cvm or $\mathrm{Cvm}_{\mathcal{C}}$. If Sub is actually a universe of embeddings, in terminology we replace the words 'subset' by 'embedding'.

In this context, a universe of subsets Lay for $\mathcal{C}$ is called a <u>universe of layered morphisms</u> (with respect to the topology) if

(2.12) for $b \in \mathrm{Mor}(\mathcal{C})$ and S a cover of cod b such that $s^{-1}b \in$ Lay for each $s \in S$, then $b \in$ Lay.

Universes of layered morphisms which satisfy the smallness condition are common.

<u>Remark 2.13:</u> The construction of C^∞-manifolds offers samples of the above items. Let **EOpen** be the class of all pairs (U,n) where $n \in \mathbb{N}$ and U is an open subset of $\mathbb{R}^n$, and let Mor(**EOpen**) be class of C^∞ functions between members of **EOpen**. In the usual way, **EOpen** becomes a category. The class of all open embeddings in **EOpen** is a universe of embeddings in the categorical sense. The classes of all C^∞ local homeomorphisms and of all C^∞ covering maps are also universes of subsets. With the standard topology, the latter becomes a universe of layered morphisms. The term 'layered' is adopted because the

more familiar word 'cover' has standard meaning in the theory of Grothendieck topologies.

Most useful categories are closed under finite inverse limits. This is not the case for **EOpen**, which is not even closed under fibered products. However, fibered products involving embeddings are well behaved. This example led to the notion of a pullback base.

By a local subset of $\mathfrak{C}$, we mean a morphism b for which there exists a cover S of dom b such that $b \circ s$ is a subset for each $s \in S$. A local subset b is called flush if there is a cover S of dom b such that $\{b \circ s : s \in S\}$ is a cover of cod b.

Proposition 2.14: Let $\mathfrak{C}$ be a topologized category. Put $\mathrm{Sub} = \mathrm{Sub}_{\mathfrak{C}}$.

(A) A $\mathfrak{C}$-isomorphism is a flush local subset.

(B) Suppose $A \in \mathfrak{C}$, $(B,b),(C,c) \in \mathfrak{C}/A$, and $(P;\pi_B,\pi_C)$ is a fibered product $(B,b)\times_A(C,c)$. If c is a local subset, then so is π_B.

(C) The composition of two local subsets is a local subset.

(D) Suppose Sub is a universe of embeddings. Suppose $A,B,C \in \mathfrak{C}$, $b \in \mathrm{Mor}_{\mathfrak{C}}(A,B)$ and $c \in \mathrm{Mor}_{\mathfrak{C}}(B,C)$. If $c \circ b$ and c are local embeddings, then so is b.

Proof: (A) and (B) are trivial.

For the rest of the argument, let $A,B,C \in \mathfrak{C}$, let $b : A \longrightarrow B$ and $c : B \longrightarrow C$ be $\mathfrak{C}$-morphisms, and let S and T be covers of A and B, respectively. Assume $c \circ t$ is a formal subset for each $t \in T$. For each $(s,t) \in S\times T$, fix a choice of $(b \circ s)\times_B t$, which we denote by $(s\times t;\pi_s,\pi_t)$.

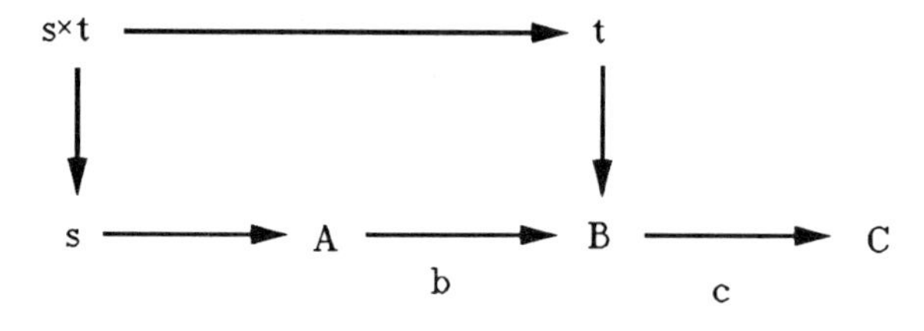

(2.15)

For $s \in S$, $t \mapsto (s\times t, \pi_s)$ is a cover of dom(s) indexed by T. Thus, $(s,t) \mapsto (s\times t, s \circ \pi_s)$ is a cover of A indexed by $S\times T$.

First, assume that $b \circ s$ is a formal subset for each $s \in S$. For $(s,t) \in S\times T$, π_t is a formal subset; consequently, $c \circ b \circ s \circ \pi_s = c \circ t \circ \pi_t$ is a formal subset. Thus, $c \circ b$ is a local subset. Next, assume that both $\{b \circ s : s \in S\}$ and $\{c \circ t : t \in T\}$ are covers. Then $s \mapsto (s \times t, \pi_t)$ is a cover for each $t \in T$. It follows that $c \circ b$ is flush.

Finally, assume Sub is a universe of embeddings, and that $c \circ b \circ s$ is an embedding for each $s \in S$. For $(s,t) \in S \times T$, $c \circ t$ and $c \circ t \circ \pi_t = c \circ b \circ s \circ \pi_s$ are embeddings, and so π_t must be an embedding. But then $b \circ s \circ \pi_s = t \circ \pi_t$ is an embedding on each $s \times t$, which imples b is a local embedding. □

§2.B Refinements and Flush Topologies

Let $\mathcal{C}$ be a category and let $A \in \mathcal{C}$. For Θ an indexed subset of $\mathcal{C}/A$ and $d : D \longrightarrow A$ a $\mathcal{C}$-morphism, we say d factors through Θ if there is $x \in \mathrm{dom}(\Theta)$ and $d' \in \mathrm{Mor}_{\mathcal{C}}(D, \mathrm{dom}\,\Theta(x))$ so that $\Theta(x) \circ d' = d$. If Θ and φ are two non-empty indexed subsets of $\mathcal{C}/A$, we say that φ factors through Θ if $\varphi(x)$ factors through Θ for each $x \in \mathrm{dom}(\varphi)$.

Let $\mathcal{C}$ be a topologized category. For $A \in \mathcal{C}$ and Θ a non-empty indexed subset of $\mathcal{C}/A$, a refinement of Θ is any indexed *cover* of A which factors through Θ. If Θ admits a refinement, we call Θ an unrefined cover of A. An unrefined cover may consist of morphisms other than formal subsets—even morphisms which are not pullbacks bases.

Proposition 2.16: Let $\mathcal{C}$ be a topologized category.

(A) A cover is an unrefined cover.

(B) Let $b \in \mathrm{Mor}(\mathcal{C})$ and let Θ be an unrefined cover of cod b. If a pullback $b^{-1}\Theta$ exists, then it is an unrefined cover.

(C) Let $A \in \mathcal{C}$ and let Θ be an unrefined cover of A. Suppose for each $i \in \mathrm{dom}(\Theta)$, φ_i is an unrefined cover $\mathrm{dom}\,\Theta(i)$. Let λ be the subdivision of Θ through φ. Then λ is an unrefined cover.

(D) Let $A \in \mathcal{C}$ and let Θ be a non-empty indexed subset of $\mathcal{C}/A$. If there is an unrefined cover of A which factors through Θ, then Θ is an unrefined cover.

Proof: (A), (B) and (D) are trivial.

(C) Let Θ^* be a refinement of Θ and, for each $i \in \mathrm{dom}(\Theta)$, let φ_i^* be a refinement of φ_i. For each $s \in \mathrm{dom}(\Theta^*)$, let

(2.17) $\quad i(s) \in \mathrm{dom}(\Theta)$ and $f_s \in \mathrm{Mor}_{\mathcal{C}/A}(\Theta^*(s), \Theta(i(s)))$.

Let

(2.18) $\Lambda^* = \{(s,t) : s \in \mathrm{dom}(\Theta^*),\ t \in \mathrm{dom}(\varphi_{i(s)}{}^*)\}$.

For each $(s,t) \in \Lambda^*$, there is a pullback $(P(s,t);\pi_s,\pi_t)$ for $\{\varphi_{i(s)}{}^*\}^{-1}f_s$; moreover, for each $s \in \mathrm{dom}(\Theta^*)$, $t \mapsto (P(s,t),\pi_s)$ is a cover of $\mathrm{dom}\,\Theta(s)$. Consequently, $(s,t) \mapsto (P(s,t),\Theta^*(s)\circ\pi_s)$ is a refinement of the given λ. □

Definition 2.19: Let $\mathcal{C}$ be a topologized category. We say $\mathcal{C}$ is flush if each unrefined cover of $\mathcal{C}$, whose members are formal subsets, is a cover.

Being flush is unimportant when a topology is used to derive cohomological information. However, the property becomes necessary in construction of global objects.

§2.C Canopies and Intrinsic Topologies

Throughout this subsection, $\mathcal{C}$ is a category.

Fix formal symbols ρ_1 and ρ_2. Let J be a be a non-empty set. Define a graph Int(J) as the set $S(J) = J \amalg J^2$ with the set of morphisms

(2.20) $M(J) = \{((i,j),\rho_1,i) : i,j \in J\} \cup \{((i,j),\rho_2,j) : i,j \in J\}$

We refer to Int(J) as the intersection graph on J. Although the above definition is proper even when J is empty, for our purposes *an intersection graph is never indexed by the empty set.* A graph F of $\mathcal{C}$-objects is called an intersection graph (over J) if it is of this type. For such an F, denote J by $\Lambda(F)$. For $i,j \in J$, we frequently denote $F[(i,j),\rho_1,i](\rho_1)$ by ρ_1 or ρ_i and $F[(i,j),\rho_2,j](\rho_2)$ by ρ_2 or ρ_j. A cone Θ of F is uniquely determined by its values on J, and is often treated as a function on J.

Let $A \in \mathcal{C}$ and let Θ is a non-empty cone into A. Put $J = \mathrm{dom}(\Theta)$, and for $j \in J$ put $A(j) = \mathrm{dom}\,\Theta(j)$. For each $(i,j) \in J^2$, suppose $A(i,j) = (A(i),\Theta(j))\times_A(A(j),\Theta(j))$ exists. Define a graph G of type Int(J) from the assignments

(2.21) $j \mapsto A(j)$ for $j \in J$,

$(i,j) \mapsto A(j,k)$ for $(j,k) \in J^2$,

$((i,j),\rho_1,i)$ goes to the first projection of $A(i,j)$, and

$((i,j),\rho_2,j)$ goes to the second projection of $A(i,j)$.

Any such G is called a canopy of Θ. The graph has an obvious cone into A, which is also referred to as its canopy, or as the canonical cone. Obviously any two canopioes of Θ are equivalent. If $b: B \longrightarrow A$ is a $\mathcal{C}$-morphism such that $b^{-1}G$ exists, then such a pullback is equivalent to a canopy of $b^{-1}\Theta$.

Definition 2.22: A non-empty indexed cone in $\mathcal{C}$ is called an intrinsic cover if its canopy exists and if the canonical cone of the canopy is a colimit. A non-empty indexed cone Θ is called an absolute cover if

(2.23.a) each value of Θ is a pullback base,

(2.23.b) every pullback of Θ is an intrinsic cover.

Obviously a pullback of an absolute cover remains an absolute cover.

Proposition 2.24: Let $A \in \mathcal{C}$ and let Θ be an indexed cone into A. Suppose φ is an absolute cover of A which factors through Θ.

(A) If the canopy of Θ exists, then Θ is an intrinsic cover.

(B) If $\mathrm{Im}(\Theta)$ consists of pullback bases, then Θ is an absolute cover.

Proof: Clearly (B) follows from (A).

(A) First, suppose Θ is an indexed cone, $A = \mathrm{cod}\,\Theta$ and φ is an absolute cover of A which factors through Θ. Let $B \in \mathcal{C}$ and $f,g \in \mathrm{Mor}_{\mathcal{C}}(A,B)$ so $f \circ \Theta(j) = g \circ \Theta(j)$ for each $j \in \mathrm{dom}(\Theta)$. Trivially, $f \circ \varphi(r) = g \circ \varphi(r)$ for $r \in \mathrm{dom}(\varphi)$, and so $f = g$.

For the rest of the argumet, assume Θ is a cone into A, G is a canopy of Θ, φ is an absolute cover of A which factors through Θ and $\beta: G \longrightarrow B$ is a cone.

Assume $r \in \mathrm{dom}(\varphi)$, $i,j \in \mathrm{dom}(\Theta)$, $f \in \mathrm{Mor}_{\mathcal{C}/A}(\varphi(r),\Theta(i))$ and $g \in \mathrm{Mor}_{\mathcal{C}/A}(\varphi(r),\Theta(j))$. Then by factoring through $G(i,j)$,

(2.25) $\quad \beta(i) \circ f = \beta(i) \circ \rho_i \circ (f \wedge g) = \beta(i,j) \circ (f \wedge g) = \beta(j) \circ g.$

Thus, there is a unique function β' on $\mathrm{dom}(\varphi)$ such that for $r \in \mathrm{dom}(\varphi)$, $i \in \mathrm{dom}(\Theta)$ and $f \in \mathrm{Mor}_{\mathcal{C}/A}(\varphi(r),\Theta(i))$, $\beta'(r) = \beta(i) \circ f$. Moreover, for $r,s \in \mathrm{dom}(\varphi)$, by choosing $i,j \in \mathrm{dom}(\Theta)$, $f \in \mathrm{Mor}_{\mathcal{C}/A}(\varphi(r),\Theta(i))$ and $g \in \mathrm{Mor}_{\mathcal{C}/A}(\varphi(s),\Theta(j))$, one derives an equality on $\varphi(r) \times_A \varphi(s)$:

(2.26) $\quad \beta'(i) \circ \pi_i = \beta(i) \circ f \circ \pi_i = \beta(i) \circ \rho_i \circ (f \times g) = \beta(i,j) \circ (f \times g) = \beta'(j) \circ \pi_j.$

Thus, β' extends to a cone on the canopy of φ. Fix $b: A \longrightarrow B$ such that $b \circ \varphi(r) = \beta'(r)$ for each $r \in \mathrm{dom}(\varphi)$.

It remains only to show that $b \circ \Theta(i) = \beta(i)$ for each $i \in \mathrm{dom}(\Theta)$. Fix $i \in \mathrm{dom}(\Theta)$. By our first remark, it suffices to show that $b\circ\Theta(i)\circ\pi_i$ equals $\beta(i)\circ\pi_i$ on $\Theta(i)\times_A\varphi(r)$ for each index r. Fix $r\in\mathrm{dom}(\varphi)$, and choose $j\in\mathrm{dom}(\Theta)$ so there is $f\in\mathrm{Mor}_{\mathcal{C}/A}(\varphi(r),\Theta(j))$. On $\Theta(i)\times_A\varphi(r)$, by factoring through $\Theta(i)\times_A\Theta(j)$,

(2.27) $$b\circ\Theta(i)\circ\pi_i = b\circ\varphi(r)\circ\pi_r = \beta(j)\circ f\circ\pi_r = \beta(j)\circ\rho_j\circ(1_{\Theta(i)}\times f) = \beta(i,j)\circ(1_{\Theta(i)}\times f)$$
$$= \beta(i)\circ\pi_i\circ(1_{\Theta(i)}\times f) = \beta(i)\circ\pi_i. \quad \square$$

Theorem 2.28: Let $\mathcal{C}$ be a category and let Sub be a universe of subsets for $\mathcal{C}$. Let Cov^{in} denote the class of all subsets of Sub which are absolute covers in $\mathcal{C}$. A Grothendieck topology Cov over Sub is said to be intrinsic if $\mathrm{Cov}\subseteq\mathrm{Cov}^{in}$. The class Cov^{in} is a Grothendieck topology, called the default topology over Sub. Moreover, an indexed subset of Sub is a cover in the default intrinsic topology if and only if it is an absolute cover.

Let $\mathcal{D}$ be a category. Let Sub be the class of pullback bases of $\mathcal{D}$. The default topology over Sub is simply called the default topology of $\mathcal{D}$.

Proof: Once we check that Cov^{in} is a topology, the final comment on covers follows from Proposition 2.24. Conditions (2.9.a,c,e) are tautological. Proposition 2.24 easy implies both (2.9.b) and (2.9.f). All that remains is the subdivision condition.

Let Θ be an absolute cover of $A\in\mathcal{C}$ and let $\varphi: r\mapsto\varphi_r$ be a subdivision function each of whose images is an absolute cover. Let λ be the subdivision through φ, and put $\Lambda=\mathrm{dom}\,\lambda$. It is trivial to show that for $B\in\mathcal{C}$ and $f,g\in\mathrm{Mor}_{\mathcal{C}}(A,B)$,

(2.29) $$(\forall k\in\Lambda,\ f\circ\lambda(k)=g\circ\lambda(k)) \quad\Rightarrow\quad f=g.$$

All that remains is the existence property of a colimit. Fix $\beta: G\longrightarrow B$ a cone from the canopy of λ.

Let $j\in\mathrm{dom}(\Theta)$. For $r,s\in\mathrm{dom}(\varphi_j)$, there is a canonical morphism $\varphi_j(r)\times_{\Theta(i)}\varphi_j(s)\longrightarrow\varphi_j(r)\times_A\varphi_j(s)$ and so, on $\varphi_j(r)\times_{\Theta(i)}\varphi_j(s)$,

(2.30) $$\beta(j,r)\circ\pi_r = \beta(j,r)\circ\pi_r\circ(1_r\times_{\Theta(j)}1_s) = \beta((j,r),(j,s))\circ(1_r\times_{\Theta(j)}1_s) = \beta(j,s)\circ\pi_s.$$

Thus, $r\mapsto\beta(j,r)$ extends to a cone on the canopy of φ_j. For $r\in\mathrm{dom}(\varphi_j)$, there is $\gamma_j\in\mathrm{Mor}_{\mathcal{D}}(\mathrm{dom}\,\Theta(j),B)$ such that $\gamma_j\circ\varphi_j(r)=\beta(j,r)$.

Obviously, it suffices to show that $j\mapsto\gamma_j$ extends to a cone on the canopy of Θ. Fix $i,j\in\mathrm{dom}(\Theta)$; we are reduced to checking that $\beta_i\circ\pi_i=\beta_j\circ\pi_j$ on $P=\Theta(i)\times_A\Theta(j)$. Pulling back φ_j along π_j further reduced the problem to showing that

(2.31) $$\gamma_i\circ\pi_i\circ(1_{\Theta(i)}\times_A\pi_j)=\gamma_i\circ\pi_i \quad\text{and}$$

$$\gamma_j \circ \pi_j \circ (1_{\Theta(i)} \times_A \varphi_j(r)) = \gamma_j \circ \varphi_j(r) \circ \pi_{j,r} = \beta(j,r) \circ \pi_{j,r}$$

agree on $\Theta(i) \times_A \lambda(j,r)$ for each $r \in \mathrm{dom}(\varphi_j)$. Fix $r \in \mathrm{dom}(\varphi_j)$. Pulling back φ_i along π_i on $\Theta(i) \times_A \lambda(j,r)$ reduces the issue to showing

(2.32) $\quad \gamma_i \circ \pi_i \circ (\varphi_i(s) \times 1_{j,r}) = \gamma_i \circ \varphi_i(s) \circ \pi_{i,s} = \beta(i,s) \circ \pi_{i,s}$ and

$$\beta(j,r) \circ \pi_r \circ (\varphi_i(s) \times 1_{j,r}) = \beta(j,r) \circ \pi_{j,r}$$

agree on $\lambda(i,s) \times_A \lambda(j,r)$ for each $s \in \mathrm{dom}(\varphi_i)$. Hypothesis on β states equality. □

Remark 2.33: Intrinsic topologies lurk behind every 'cut-and-paste' construction. Intuitively, all properties—especially mapping properties—of an object built from charts are derived the objects which make up the charts. The formal expression of this idea is to require that the pasted object be a colimit of its charts. *The nature of the process fixes the topology; essentially, we always begin and end with an intrinsic topology.* From this perspective, the standard and étale topologies on the category of rings differ because the they begin with *different kinds of formal subsets.* The notion of 'cover' is really the same for both!

Our objective is to derive, from an initial $\mathfrak{C}$ with Grothendieck topology, another topologized category $\mathfrak{C}^+$ filled with 'locally $\mathfrak{C}$' objects. From our viewpoint, the choice of formal subsets essentially determines the topology. Topologizing $\mathfrak{C}^+$ is a question of recognizing when an added global morphism should be a formal subset. The reader will find that our emphasis throughout this work is on the class of formal subsets, rather than on the choice of covers.

§2.D Covering/Local Criterion for Subsets

Let $\mathfrak{C}$ be a topologized category. A $\mathfrak{C}$-morphism b is said to meet the covering/local criterion for subsets, or CLCS, if there is a cover S of dom b such that for each $s \in S$,

(2.34.a) $b \circ s$ is a formal subset,

(2.34.b) a pullback of $(b \circ s)$ along b exists, and its projection to dom s is a covering morphism.

Roughly, (2.34.a,b) means that b is a local subset and projections $b \times_{\mathrm{cod}\, b} b \longrightarrow \mathrm{dom}\, b$ are covering morphisms. Obviously (2.34.a,b) is equivalent to existence of covers S_a and S_b of dom b which separately satisfy the respective conditions (2.34.a) and (2.34.b). Observe that if s is a morphism into dom b and the projection $b \times_{\text{cod}\, b} (b \circ s) \longrightarrow \text{dom}\, s$ is a formal subset, then Proposition 2.10 states that projection is a covering morphism.

The CLCS is part of a solution rather than an initial definition. Again, imagine an topologized category $\mathfrak{C}$ is embedded in a larger category $\mathfrak{C}^+$. We wish to induce a Grothendieck topology on $\mathfrak{C}^+$; as already remarked, this amounts to deciding which $\mathfrak{C}^+$-morphisms should be regarded as formal subsets. Consider two variations on this problem.

(2.35.a) First, there is the theme of this paper, namely passing from local structure to universe of global objects. The global objects should inherit a canonical topology.

(2.35.b) Recall construction of the spectrum of a ring, a step in building schemes. One begins with a category $\mathfrak{C}$ of rings and the class Sub of every morphism which is a localization with respect to one element. All foundational lemmas are performed with these localizations serving for embeddings. Yet there are ring homomorphisms which are open embeddings of schemes but which are not localizations. The latter are morphisms which meet (2.34.a,b) with respect to Sub. When passing from rings to schemes, mathematical pressures require one to broaden the initial choice of subset.

In either situation, the concept of a *local* subset extends easily. Each $\mathfrak{C}^+$ object comes with canonical charts; a morphism is regarded as a local subset if each restriction to a chart is a local subset in $\mathfrak{C}$. Still, classical categories offers examples of local subsets which are not subsets. The finite covering map of point set topology is archetypical.

When $\mathfrak{C}^+$ is a category of sheaves over topological spaces, there is a topological criterion. A morphism b is an embedding if each x in the image of the *base map* of b has an open neighborhood U such that b, restricted to the sheaf over $b^{-1}U$, yields an isomorphism. The CLCS is an effort to translate this condition to category theory. Assuming b is a local subset, we replace open subsets in the base space by morphisms $b \circ s$ where s is a formal subset into dom b and $b \circ s$ is also a formal subset. The sheaf pullback is replaced by categorical pullback; however, covering morphisms are substituted for isomorphisms, as the latter constraint is too restrictive for the étale topology.

Condition (2.34.b) is pointless as a guideline unless we develop methods for lifting the notion of 'Cvm-morphism'. The latter, no easy feat, is discussed later.

It is simple to modify an existing Grothendieck topology by changing the notion of covering morphism.

Proposition 2.36: Let ($\mathfrak{C}$,Sub,Cov) be a topologized category. Let Lay be a universe of layered morphisms for $\mathfrak{C}$. Assume each Lay-morphism is a covering morphism with respect to Cov. Define Sub_1 to be the class of Sub-morphisms b such that both projections $b\times_{\mathrm{cod}\,b} b \longrightarrow \mathrm{dom}\, b$ are Lay-morphisms. Let Cov_1 be the class of covers consisting of Sub_1-morphisms.

(A) Sub_1 is a universe of subsets for $\mathfrak{C}$ and Cov_1 is a Grothendieck topology for Sub_1. We call (Sub_1,Cov_1) the cover reduction of $\mathfrak{C}$ through Lay.

(B) Lay is the class of covering morphisms of the topology Cov_1.

(C) If Cov is flush, than so is Cov_1.

(D) If Cov is intrinsic, then so is Cov_1.

(E) If Sub and Lay are universes of embeddings, then Sub_1 is a universe of embeddings.

Proof: Every $\mathfrak{C}$-isomorphism is a Lay-morphism. Thus, for $b \in \mathrm{Mor}(\mathfrak{C})$, both projections $b\times_{\mathrm{cod}\,b} b \longrightarrow \mathrm{dom}\, b$ are Lay-morphisms if and only if either is a Lay-morphism.

From Proposition 1.10 and Proposition 1.19, the pullback of a Sub_1-morphism is also in Sub_1. Consider the diagram

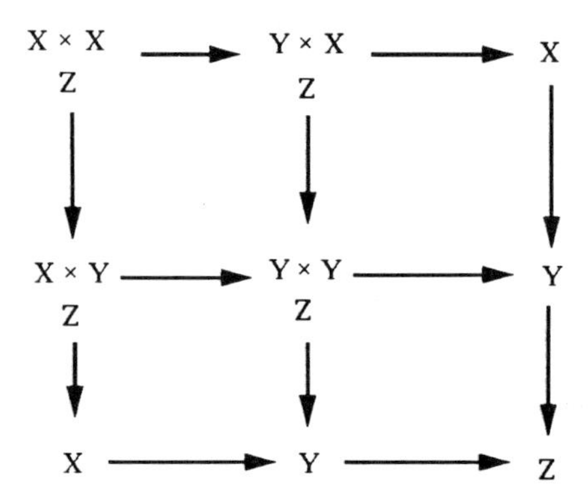

(2.37)

where $b: Y \longrightarrow Z$ and $c: X \longrightarrow Y$ are Sub_1-morphisms. Then $c\times_Z 1_Y$ is a Sub_1-morphism. By Corollary 1.15, $c\times_Z 1_X$ is a Lay-morphism. By assumption, projection $Y\times_Z Y \longrightarrow Y$ is a Lay-morphism, which implies that projection $Y\times_Z X \longrightarrow X$ is as well. Composing two layered morphisms yields a member of Lay, so the first projection $X\times_Z X \longrightarrow X$ is a Lay-morphism. Hence, Sub_1 is a universe of subsets.

Obviously Cov_1 is a topology over Sub_1. (C) and (D) are immediate.

(B) Suppose $b: B \longrightarrow A$ is a covering morphism with respect to Cov_1. Projection $B\times_A B \longrightarrow B$ can be regarded as a pullback of b along a covering morphism. Condition (2.12) implies $b \in Lay$.

(E) Assume both Lay and Sub are universes of embeddings. Let $b: Y \longrightarrow Z$ and $c: X \longrightarrow Y$ be $\mathfrak{C}$-morphisms such that $b, b\circ c \in Sub_1$. Consider diagram (2.37) again, this time with property (2.4) in mind. We are given that the projections of $X\times_Z X \longrightarrow X$ and $Y\times_Z Y \longrightarrow Y$ are in Lay. Now projections $Y\times_Z X \longrightarrow X$ and $X\times_Z Y \longrightarrow X$ are pullbacks of the latter, and so are Lay-morphisms. Property (2.4) for Lay now yields that $1_X\times c, c\times 1_X \in Lay$. These two morphisms may be considered as projections of $(1_Y\times c)\times_{(Y\times_Z Y)}(1_Y\times c)$.

Because Sub is a universe of embeddings, the hypothesis implies that $c \in Sub$. Consequently, $1_Y\times c$, a pullback of c, is also in Sub. The previous papragraph now implies that $1_Y\times c \in Sub_1$.

There is a Lay-morphism along which the pullback of c is in Sub_1. Hence, there is a Lay-morphism into X along which pullback of either projection $c\times_Y c \longrightarrow X$ is a Lay-morphism. By (2.12), these two projections must be Lay-morphisms. □

<u>Definition 2.38:</u> Let $(\mathfrak{C}, Sub, Cov)$ be a topologized category. Put $Cvm = Cvm_{\mathfrak{C}}$. We say that $\mathfrak{C}$ meets the <u>CLCS condition</u> if

(2.39.a) Cvm is a universe of layered morphisms with respect to (Sub,Cov),

(2.39.b) every $\mathfrak{C}$-morphism which meets the CLCS is a formal subset.

<u>Corollary 2.40:</u> Let $(\mathfrak{C}, Sub, Cov)$ be a topologized category. Let Lay be a universe of layered morphisms for $\mathfrak{C}$. Assume each Lay morphism is a covering morphism with respect to Cov. Let (Sub_1, Cov_1) be the cover reduction through Lay. If Cov satisfies the CLCS condition, then so does Cov_1.

<u>Proof:</u> Suppose $b \in Mor(\mathfrak{C})$ meets the CLCS condition with respect to Cov_1. Then $b \in Sub$. It follows that $b \in Cov_1$. □

§2.E Continuous Functors

There are two archetypes for functors on categories with Grothendieck topology.

(2.41.a) For $f : X \longrightarrow Y$ a continuous function between topological spaces, the pullback of open subsets yields a functor from the (small) category of open subsets of Y into that of X.

(2.41.b) If $\mathcal{C}$ is any category consisting of sheaves over topological bases, then there is a functor which to each object assigns the space of global sections.

Intuitively, a function f as in (2.41.a) should begat a universal functor if and only if the topology on X is the pullback topology along f. This idea is difficult to formalize, and led the author to the CLCS conditions.

Definition 2.42: Let $\mathcal{C}$ be a topologized category, let $\mathcal{D}$ be a category, and let $\Gamma : \mathcal{C} \longrightarrow \mathcal{D}$ be a covariant functor. We say Γ is a functor of sections if for $X \in \mathcal{C}$, Θ an indexed cover of X and $\Theta^{\#} : X_0 \longrightarrow X$ the cone of the canopy of Θ, $\Gamma(\Theta^{\#})$ is a colimit.

In the rest of this definition, assume $\Gamma : \mathcal{C} \longrightarrow \mathcal{D}$ is a covariant functor between two categories with Grothendieck topologies.

We say Γ is continuous if

(2.43.a) Γ maps formal $\mathcal{C}$-subsets to formal $\mathcal{D}$-subsets,

(2.43.b) Γ maps $\mathcal{C}$-covers to $\mathcal{D}$-covers,

(2.43.c) if $A \in \mathcal{C}$ and $(B,b),(C,c) \in \mathcal{C}/A$ such that b is a formal subset, then Γ preserves the fibered product $(B,b)\times_A(C,c)$.

Note that a continuous functor sends covering morphisms to covering morphisms.

Let $\Gamma : \mathcal{C} \longrightarrow \mathcal{D}$ be a continuous functor. For $D \in \mathcal{D}$, define a cover of D through Γ to be a function $\Theta : j \longmapsto (B(j),b(j))$ such that

(2.44.a) $B(j) \in \mathcal{C}$ and $b(j) \in \mathrm{Sub}_{\mathcal{D}}(\Gamma(B(j)),D)$ for each $j \in \mathrm{dom}(\Theta)$,

(2.44.b) $j \longmapsto (\Gamma(B(j)),b(j))$ is an $\mathcal{D}$-cover of D indexed by $\mathrm{dom}(\Theta)$.

We frequently identify Θ with the cover of (2.44.b).

Suppose Γ is continuous. Assume

(2.45.a) for $A,B \in \mathcal{C}$, the restriction of Γ determines a bijection $\mathrm{Mor}_{\mathcal{C}}(A,B) \longrightarrow \mathrm{Mor}_{\mathcal{D}}(\Gamma(A),\Gamma(B))$,

(2.45.b) for $b \in \mathrm{Mor}(\mathcal{C})$, if $\Gamma(b)$ is a formal $\mathcal{D}$-subset, then b is a formal $\mathcal{C}$-subset,

(2.45.c) for S a non-empty cone of $\mathcal{C}$, if $\Gamma(S)$ is a $\mathcal{D}$-cover then S is a $\mathcal{C}$-cover.

(2.45.d) each $D \in \mathcal{D}$ admits a cover through Γ.

For convenience, call Γ a <u>weak functorial embedding</u>. In a classical context, the functor from a local category to one of global objects meets these conditions. However, the list is not quite long enough to be useful (hence the adjective 'weak').

Let $\Gamma : \mathcal{C} \longrightarrow \mathcal{D}$ be a weak functorial embedding. Suppose Θ is a class-theoretic function which to each $D \in \mathcal{D}$ assigns a cover of D through Γ. There may be *several* topologies of $\mathcal{D}$ with respect to which Γ is a weak functorial embedding and with respect to which $\Theta(D)$ is a cover of D through Γ for each D. Again, this ambiguity is related to the formulation of the CLCS criterion. We say Γ is a <u>CLCS functorial embedding</u> if for $A \in \mathcal{C}$ and $d : D \longrightarrow \Gamma(A)$ a covering morphism in $\mathcal{D}$, there is a covering morphism $b : B \longrightarrow A$ in $\mathcal{C}$ such that (D,d) is $\mathcal{D}/\Gamma(A)$-isomorphic to $(\Gamma(B),\Gamma(b))$.

After a simple lemma, we get a uniqueness statement for CLCS functorial embeddings.

<u>Lemma 2.46:</u> Suppose $\mathcal{D}$ is a topologized category which meets the CLCS condition. Suppose $b : B \longrightarrow A$ is a $\mathcal{D}$-morphism and Θ is an indexed cover of A. If for each $j \in \mathrm{dom}(\Theta)$ the pullback of b along $\Theta(j)$ is a formal subset, then b is a formal subset.

<u>Proof:</u> Obviously b is a local subset. It suffices to show that for each $j \in \mathrm{dom}(\Theta)$, $b \times_A (b \circ \{b^{-1}\Theta(j)\}) \longrightarrow \mathrm{dom}\, b^{-1}\Theta(j)$ exists and is a formal subset. Fix $j \in \mathrm{dom}(\Theta)$. Since $b \circ \{b^{-1}\Theta(j)\} = \Theta(j) \circ b_|$, this morphism is a formal subset, and all pullbacks along it exist. Using the same equality, Proposition 1.10 identifies the projection with a pullback of $\Theta(j)^{-1}b$, which is given as a formal subset. □

<u>Proposition 2.47:</u> Let $\mathcal{C}$ be a topologized category, let $\mathcal{D}$ be a category, and let $\Gamma : \mathcal{C} \longrightarrow \mathcal{D}$ be a covariant functor. Suppose Θ is a class-theoretic function which to each $D \in \mathcal{D}$ assigns a function $\Theta_D : j \mapsto (B(D,j), b(D,j))$ such that

(2.48) $B(D,j) \in \mathcal{C}$ and $b(D,j) \in \mathrm{Mor}_{\mathcal{D}}(\Gamma(B(j)),D)$ for each $j \in \mathrm{dom}(\Theta_D)$.

(A) There is at most one universe of subsets Sub for $\mathcal{D}$ for which there exists a Grothendieck topology Cov such that

(2.49.a) Cov meets the CLCS condition,

(2.49.b) Γ is a CLCS functorial embedding with respect to (Sub,Cov),

(2.49.c) for $D \in \mathcal{D}$, θ_D is a cover of D through Γ.

(B) There is at most one flush Grothendieck topology (Sub,Cov) for $\mathcal{D}$ which meets (2.49.a,b,c).

Proof: Suppose (Sub,Cov) is a topology on $\mathcal{D}$ which meets (2.49.a,b,c).

(A) It suffices to characterize members of Sub in term of the hypothesized data. By condition (2.49.a), Sub can be characterized provided we first characterize covering morphisms and local subsets for $\mathcal{D}$.

Suppose $c: C \longrightarrow A$ is a $\mathcal{D}$-morphism. Then c is a covering morphism with respect to (Sub,Cov) if and only if for each $j \in \mathrm{dom}(\theta_A)$, $b(A,j)^{-1}c$ is a covering morphism with respect to Cov. Since $\mathrm{dom}\, b(A,j) = \Gamma(B(A,j))$ and Γ is a CLCS functorial embedding, the latter statement is equivalent to requiring that each pullback be isomorphic to a covering morphism under Γ. Thus, covering morphisms of (Sub,Cov) are uniquely determined.

Let $c: C \longrightarrow A$ be a $\mathcal{D}$-morphism. From (2.45.a,b,c) and Lemma 2.46, it is simple to show that

(2.50.a) c is a local subset with respect to (Sub,Cov) if and only if for each $j \in \mathrm{dom}(\theta_C)$ there is a cover φ_j of B(C,j) in $\mathcal{C}$ such that for $x \in \mathrm{dom}(\varphi_j)$, $c \circ b(C,j) \circ \Gamma(\varphi_j(x)) \in \mathrm{Sub}$,

(2.50.b) $c \in \mathrm{Sub}$ if and only if the pullback of c along $\theta_A(k)$ is in Sub for each $k \in \mathrm{dom}(\theta_A)$.

To recognize when c is a formal subset in (Sub,Cov), it suffices to determine which $\mathcal{D}$-morphisms into images under Γ are in Sub. The previous paragraph reduces the latter question to recognizing when a $\mathcal{D}$-morphism into an image under Γ is a local subset. But (2.50.a) and conditions (2.45.a,b,c) provide a description which depends solely on the topology of $\mathcal{C}$. Thus, Sub is uniquely determined.

(B) We must characterize Cov in terms of Sub and the given data, using the added assumption that Cov is flush. If $A \in \mathcal{D}$ and φ is a cone of Sub-morphisms into A, then flushness of $\mathcal{D}$ implies that φ is an indexed cover if and only if for each $j \in \mathrm{dom}(\theta_A)$, $b(A,j)^{-1}\varphi$ is a cover. Now suppose $A \in \mathcal{C}$ and φ is a cone of Sub-morphisms into $\Gamma(A)$. Again, flushness implies that φ is a cover if and only if its subdivision by $x \longmapsto \theta_{\mathrm{dom}\, \varphi(x)}$ is a cover. The issue is reduced to questions of $\mathcal{C}$ and Γ. □

Remark 2.51: Each of the functorial properties defined in this section is preserved under functorial equivalence.

PART II CANOPIES

Let $\mathfrak{C}$ be a category with Grothendieck topology. An abstract canopy (defined below) serves several roles. Formally, it is a graph; intuitively, it is a graph derived from a cover of a 'locally $\mathfrak{C}$' object under construction. The class of canopies, along with a suitable concept of morphism, create a new category of global objects. However, morphisms are difficult to describe.

Imagine two objects A_1, A_2 which lie in a category containing $\mathfrak{C}$, and suppose each A_i is assigned a cover Θ_i whose canopy G_i consists only of $\mathfrak{C}$-objects. The problem is to describe a morphism $f: A_1 \longrightarrow A_2$ in terms of graph data. If for each $j \in \mathrm{dom}(\Theta_1)$ there is $\sigma(j) \in \mathrm{dom}(\Theta_2)$ and $f(j) \in \mathrm{Mor}_{\mathfrak{C}}(G_1(j), G_2(\sigma(j))$ so $f \circ \Theta_1(j) = \Theta_2(\sigma(j)) \circ f(j)$, then f can be represented by the assignment $j \mapsto (f(j), \sigma(j))$. It is not difficult to determine when two such assignments represent the same morphism. Alas, not every morphism allows for such an expression. In general, for a given f, there is such a representation with respect to a *subdivision* of the cover Θ_1. This idea cannot be exploited until the intuition of refinement can be made rigorous.

The author proceeds in stages. First, we create a category $\mathfrak{C}^p$ of canopies in which morphisms are of the simplest kind—that is, a $\mathfrak{C}^p$-morphism is an equivalence class of indexed families of $\mathfrak{C}$-morphisms. This category is too restricted. However, a Grothendieck topology can be granted to $\mathfrak{C}^p$. We have already introduced formal notions for subdivision and refinement. Simple as the definitions are, their bases in a categorical topology assure that they behave according to our intuition. Later, we build on the concepts to enlarge the class of morphisms in $\mathfrak{C}^p$. The latter process is the *smoothing* functor.

§3 The Category of Canopies

For this section, assume

(3.1) $(\mathfrak{C}, \mathrm{Sub}, \mathrm{Cov})$ is a category with Grothendieck topology.

The first step is to generalize the notion of canopy introduced in Section 2.

Let F be an intersection graph over $\mathfrak{C}$. For $i \in \Lambda(F)$, a morphism $\delta: F(i) \longrightarrow F(i,i)$ such that $\rho_1 \circ \delta = \rho_2 \circ \delta = 1_{F(i)}$ is called a <u>diagonal at i</u> (for F). Now let $(i,j,k) \in \Lambda(F)^3$, and suppose

(3.2) $(P; \pi_i, \pi_k)$ is a fibered product for $(F(i,j), \rho_j) \times_{F(j)} (F(j,k), \rho_j)$.

A morphism $\omega: P \longrightarrow F(i,k)$ is called a <u>transition morphism</u> (for (i,j,k) and F) if

(3.3.a) $(P;\omega,\pi_i)$ is a fibered product for $(F(i,k),\rho_i) \times_{F(i)} (F(i,j),\rho_i)$.

(3.3.b) $(P;\omega,\pi_k)$ is a fibered product for $(F(i,k),\rho_k) \times_{F(k)} (F(j,k),\rho_k)$.

Definition 3.4: An intersection graph F is a canopy with respect to ($\mathcal{C}$,Sub,Cov) if the following conditions hold:

(3.5.a) For $i,j \in \Lambda(F)$, $\rho_i : F(i,j) \longrightarrow F(i)$ and $\rho_j : F(i,j) \longrightarrow F(j)$ are formal subsets.

(3.5.b) For $i,j \in \Lambda(F)$, $X \in \mathcal{C}$ and $f,g \in Mor_{\mathcal{C}}(X,F(i,j))$, if $\rho_i \circ f = \rho_i \circ g$ and $\rho_j \circ f = \rho_j \circ g$, then $f=g$.

(3.5.c) For $i \in \Lambda(F)$, there is a diagonal at i for F.

(3.5.d) For $i,j \in \Lambda(F)$, there is a $\mathcal{C}$-morphism $\varphi : F(i,j) \longrightarrow F(j,i)$ such that $\rho_i \circ \varphi = \rho_i$ and $\rho_j \circ \varphi = \rho_j$.

(3.5.e) There exists a transition morphism for each $(i,j,k) \in \Lambda(F)^3$.

The five conditions of Definition 3.4 imply stronger constraints. Condition (3.5.b) assures there is a unique diagonal at each $i \in \Lambda(F)$. The morphism φ of (3.5.d) is both uniquely defined and an isomorphism. Condition (3.5.a) assures existence of the fibered product upon which the transition morphism of (3.5.e) is to be defined; again, the transition morphism of a triple (i,j,k) is unique. For $i \in \Lambda(F)$, Proposition 2.10 implies that each of the morphisms ρ_1 and ρ_2 on F(i,i) is a covering morphism.

The canopy of an indexed family of formal subsets is a canopy in the above sense. Intuitively, a canopy F is a 'new' object for $\mathcal{C}$ with an assigned cover $\{\iota_j : F(j) \longrightarrow F\}_{j \in \Lambda(F)}$. Each triple $(F(j,k),\rho_j,\rho_k)$ should be the fibered product $F(j) \times_F F(k)$.

Let Can($\mathcal{C}$), or Can, denote the class of canopies over $\mathcal{C}$.

Proposition 3.6: Let $B \in \mathcal{C}$ and let F be a canopy. By a chart map from $B \longrightarrow F$, we mean a pair (b,j) where $j \in \Lambda(F)$ and $b \in Mor_{\mathcal{C}}(B,F(j))$. The set of chart maps from B to F is denoted Ch(B,F). Let

(3.7) $$R = \{((b,j),(c,k)) \in Ch(B,F) \;:\; \exists d \in Mor_{\mathcal{C}}(B,F(j,k)) \text{ so } b=\rho_j \circ d \text{ and } c=\rho_k \circ d \}.$$

Then R is an equivalence relation on Ch(B,F).

The set of R-equivalence classes is denoted by Cc(B,F). For $(b,j),(c,k) \in Ch(B,F)$, write $(b,j) \sim (c,k)$ if $((b,j),(c,k)) \in R$. For $(b,j) \in Ch(B,F)$, the class of (b,j) is called its chart class and is denoted by $[b,j]$ or $[b]$.

In what follows, fix $B \in \mathcal{C}$ and $F \in \mathrm{Can}(\mathcal{C})$.

(A) Suppose $D \in \mathcal{C}$, $d \in \mathrm{Mor}_{\mathcal{C}}(D,B)$ and $\beta \in \mathrm{Cc}(B,F)$. Then there is a unique $\delta \in \mathrm{Cc}(D,F)$, denoted by $\beta \circ d$, such that for each $(b,j) \in \alpha$, $(b \circ d, j) \in \delta$.

(B) Suppose $D,E \in \mathcal{C}$, $d \in \mathrm{Mor}_{\mathcal{C}}(D,B)$, $e \in \mathrm{Mor}_{\mathcal{C}}(E,D)$ and $\beta \in \mathrm{Cc}(B,F)$. Then $(\beta \circ d) \circ e = \beta \circ (d \circ e)$.

(C) For $\beta \in \mathrm{Cc}(B,F)$, $\beta \circ 1_B = \beta$.

(D) Assume Sub is a universe of embeddings. Suppose $(b,j),(c,k) \in \mathrm{Ch}(B,F)$ and $(b,j) \sim (c,k)$. Then $b \in \mathrm{Sub}(B,F(j))$ if and only if $c \in \mathrm{Sub}(B,F(k))$.

Proof: First, fix B and F. Conditions (3.5.c,d,e) easily imply that R is an equivalence relation.

(A) is simple. Statements (B) and (C) are tautological corollaries to (A).

(D) If $j,k \in \Lambda(F)$ and $b: B \longrightarrow F(j,k)$ so that $\rho_j \circ b$ is an embedding, then condition (2.4) implies that b is an embedding. □

Definition 3.8: Let F and G be canopies. By a filtered map from F to G, we mean a function Θ which assigns to each $j \in \Lambda(F)$ an element $\Theta(j) \in \mathrm{Cc}(F(j),G)$ such that for $j,k \in \Lambda(F)$, $\Theta(j) \circ \rho_j = \Theta(k) \circ \rho_k$ on $F(j,k)$. A function which assigns to each $j \in \Lambda(F)$ an element in $\mathrm{Ch}(F(j),G)$ is called a half-map. A half-map Θ_0 is said to represent a filtered map Θ (or is a representative of Θ) if $\Theta(j) = [\Theta_0(j)]$ for each $j \in \Lambda(F)$.

The set of filtered maps F to G is denoted by $\mathrm{Mor}_{\mathrm{Can}}(F,G)$. For $F \in \mathrm{Can}$, $j \longmapsto [1_{F(j)}, j]$ is a filtered map from F to F which we denote by $\mathbb{1}_F$.

Lemma 3.9: Let $F,G,H \in \mathrm{Can}$, let $\Theta \in \mathrm{Mor}_{\mathrm{Can}}(F,G)$ and let $\varphi \in \mathrm{Mor}_{\mathrm{Can}}(G,H)$. Then there is a unique $\beta \in \mathrm{Mor}_{\mathrm{Can}}(F,H)$ with the property that if

(3.10) $\quad x \longmapsto (\Theta_0(x), i(x))$ is a representative of Θ, and

$\qquad y \longmapsto (\varphi_0(y), j(y))$ is a representative of φ,

then

(3.11) $\quad x \longmapsto (\varphi_0(i(x)) \circ \Theta_0(x), j(i(x)))$

represents β. We denote β by $\varphi \circ \Theta$.

<u>Proof:</u> If Θ^0 and φ^0 are half-maps of the form (3.10), then denote the half-map of (3.11) by $\varphi^0 \circ \Theta^0$.

First, suppose $\Theta^0 : x \mapsto (\Theta_0(x), i(x))$ and $\varphi^0 : y \mapsto (\varphi_0(y), j(y))$ represent Θ and φ, respectively. Fix $u,v \in \Lambda(F)$. Observe that

(3.12) $$[(\varphi^0 \circ \Theta^0)(u)] \circ \rho_u = [\varphi_0(i(u)) \circ \Theta_0(u) \circ \rho_u, j(i(u))],$$
$$[(\varphi^0 \circ \Theta^0)(v)] \circ \rho_v = [\varphi_0(i(v)) \circ \Theta_0(v) \circ \rho_v, j(i(v))].$$

Since Θ is a filtered map, there is $f : F(u,v) \longrightarrow G(i(u),i(v))$ such that

(3.13) $$\Theta_0(u) \circ \rho_u = \rho_{i(u)} \circ f \quad \text{and} \quad \Theta_0(v) \circ \rho_v = \rho_{i(v)} \circ f.$$

Since φ is a filtered map, there is $g : G(i(u),i(v)) \longrightarrow H(j(i(u)),j(i(v)))$ such that

(3.14) $$\varphi_0(i(u)) \circ \rho_{i(u)} = \rho_{j(i(u))} \circ g \quad \text{and} \quad \varphi_0(i(v)) \circ \rho_{i(v)} = \rho_{j(i(v))} \circ g.$$

Then

(3.15) $$\rho_{j(i(u))} \circ g \circ f = \varphi_0(i(u)) \circ \Theta_0(u) \circ \rho_u \quad \text{and}$$
$$\rho_{j(i(v))} \circ g \circ f = \varphi_0(i(v)) \circ \Theta_0(v) \circ \rho_v.$$

Thus, $\varphi^0 \circ \Theta^0$ does represent a filtered map.

Next, suppose $\Theta^0 : x \mapsto (\Theta_0(x), i(x))$ represents Θ, and $\varphi^0 : y \mapsto (\varphi_0(y), j_0(y))$ and $\varphi^1 : y \mapsto (\varphi_1(y), j_1(y))$ each represent φ. By Proposition 3.6, for $u \in \Lambda(F)$.

(3.16) $$[(\varphi^0 \circ \Theta^0)(u)] = [\varphi_0(i(u)), j_0(i(u))] \circ \Theta_0(u)$$
$$= [\varphi_1(i(u)), j_1(i(u))] \circ \Theta_0(u) = [(\varphi^1 \circ \Theta^0)(u)].$$

Finally, suppose $\Theta^0 : x \mapsto (\Theta_0(x), i_0(x))$ and $\Theta^1 : x \mapsto (\Theta_1(x), i_1(x))$ represent Θ, and $\varphi^0 : y \mapsto (\varphi_0(y), j(y))$ represents φ. Fix $u \in \Lambda(F)$, and take $f : F(u) \longrightarrow G(i_0(u), i_1(u))$ so

(3.17) $$\Theta_0(u) = \rho_{i_0(u)} \circ f \quad \text{and} \quad \Theta_1(u) = \rho_{i_1(u)} \circ f.$$

Take $g : G(i_0(u), i_1(u)) \longrightarrow H(j(i_0(u)), j(i_1(u)))$ so that

(3.18) $$\varphi_0(i_0(u)) \circ \rho_{i_0(u)} = \rho_{j(i_0(u))} \circ g \quad \text{and}$$
$$\varphi_0(i_1(u)) \circ \rho_{i_1(u)} = \rho_{j(i_1(u))} \circ g.$$

Then

(3.19) $$(\varphi^0 \circ \Theta^0)(u) = \rho_{j(i_0(u))} \circ g \circ f \quad \text{and} \quad (\varphi^0 \circ \Theta^1)(u) = \rho_{j(i_1(u))} \circ g \circ f$$
$$\Rightarrow \quad [(\varphi^0 \circ \Theta^0)(u)] = [(\varphi^0 \circ \Theta^1)(u)].$$

The lemma follows directly. □

<u>Definition 3.20:</u> The following data determines a category:

(3.21.a) the object class Can,

(3.21.b) the morphism assignment $(A_0,B_0)\mapsto \mathrm{Mor}_{\mathrm{Can}}(A_0,B_0)$ on Can^2,

(3.21.c) the composition assignment which maps each triple $(A_0,B_0,C_0)\in \mathrm{Can}^3$ to

$$(c_0,b_0)\longmapsto c_0\circ b_0 \quad \text{on } \mathrm{Mor}_{\mathrm{Can}}(B_0,C_0)\times \mathrm{Mor}_{\mathrm{Can}}(A_0,B_0)\longrightarrow \mathrm{Mor}_{\mathrm{Can}}(A_0,C_0),$$

(3.21.d) the identity assignment $A_0\mapsto \mathbb{1}_{A_0}\in \mathrm{Mor}_{\mathrm{Can}}(A_0,A_0)$.

This category is denoted by Can, Can($\mathfrak{C}$,Sub,Cov), or (discussed later) by $\mathfrak{C}^p$.

For each $A\in\mathfrak{C}$, let A^p denote the canopy determined by

(3.22.a) $\Lambda(A^p)=\{1\}$,

(3.22.b) $A^p(1)=A$ and $A^p(1,1)=A$,

(3.22.c) on $A^p(1,1)$, $\rho_1=\rho_2=1_A$.

For $b:A\longrightarrow B$ a $\mathfrak{C}$-morphism, let b^p denote the Can-morphism $\{(1,[b,1])\}$. The following remarks are elementary:

(A) For $A,B\in\mathfrak{C}$, the function $b\mapsto b^p$ is a bijection $\mathrm{Mor}_{\mathfrak{C}}(B,A)\longrightarrow \mathrm{Mor}_{\mathrm{Can}}(B^p,A^p)$.

(B) For $A\in\mathfrak{C}$ and $B_0\in\mathrm{Can}$, the function $b_0\mapsto b_0(1)$ determines a bijection $\mathrm{Mor}_{\mathrm{Can}}(A^p,B_0)\longrightarrow \mathrm{Cc}(A,B_0)$. We freely identify these spaces in this manner.

(C) The class-theoretic functions $A\mapsto A^p, b\mapsto b^p$ determines a covariant functor from $\mathfrak{C}\longrightarrow\mathrm{Can}$. We refer to it as the <u>pasting</u> functor. An object in Can which is isomorphic to A^p for some $A\in\mathfrak{C}$ is called <u>affine</u>.

A morphism from an affine object into any canopy is called affine. A set or indexed family of affine morphisms is also referred to as affine.

<u>Theorem 3.23:</u> Let F be a canopy. For each $j\in\Lambda(F)$, we refer to $[1_{F(j)},j]\in \mathrm{Cc}(F(j),F)$—or to its interpretation as a morphism from $F(j)^p$ to F—by ι_j, the <u>j-th coordinate chart</u> of F. If $A\in\mathfrak{C}$, $B_0\in\mathrm{Can}$ and $(b,j)\in\mathrm{Ch}(A,B_0)$, then $[b,j]=\iota_j\circ b^p$ as morphisms on A^p.

Let $B_0\in\mathrm{Can}$. Now B_0 can be regarded as a graph in $\mathfrak{C}^p$; let B_1 denote its image under the pasting functor, regarded as a graph in $\mathfrak{C}^p$. Let $\iota^{\#}:B_1\longrightarrow B_0$ be the unique cone such that $\iota(j)=\iota_j$ for each $j\in\Lambda(F)$. Then $\iota^{\#}$ is a colimit in the category Can.

<u>Proof:</u> Tautological. □

<u>Corollary 3.24:</u> Suppose G is a graph of Can-objects, and let $\Theta: B_0\longrightarrow G$ be a source. Suppose for each $A\in\mathfrak{C}$ and each source $\varphi: A^p\longrightarrow G$, there is a unique factoring of φ

through Θ. Then Θ is an inverse limit.

Proof: Once Theorem 3.23 is established, this is a tautological consequence of Definition 3.20.B. □

Corollary 3.25: Suppose G is a graph of $\mathfrak{C}$-objects, and let $\Theta : B \longrightarrow G$ be a source. Let G^p and Θ^p denote the images of the graph G and the source Θ, respectively, under the pasting functor. If Θ is an inverse limit in $\mathfrak{C}$, then Θ^p is an inverse limit in Can.

Proof: Once Corollary 3.24 and part (A) of Definition 3.20 are accepted, the claim is trivial. □

To fully exploit Can, we need fibered products.

Let ρ be a formal symbol. Let S be the set $\{1,2,3,(1,2),(2,3)\}$, and put

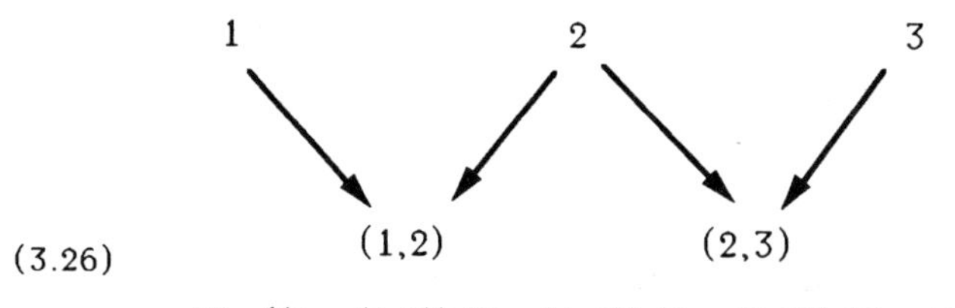

(3.26)

$$M = \{(1,\rho,(1,2)),(2,\rho,(1,2)),(2,\rho,(2,3)),(3,\rho,(2,3))\}$$

We call (S,M) the W-graph. If F is a graph of objects of type W, we refer to $F[1,(1,2)](\rho)$ or $F[2,(1,2)](\rho)$ by ρ or ρ_{12}, and to $F[2,(2,3)](\rho)$ or $F[3,(2,3)](\rho)$ by ρ or ρ_{23}. A source of a W-graph is indicated by its morphisms to the 1, 2 and 3 objects.

Proposition 3.27: Let $\mathcal{D}$ be a category and let G be a W-graph of $\mathcal{D}$-objects. Suppose

(3.28.a) $(P_{12};\pi_1,\pi_2)$ is a fibered product $(G(1),\rho)\times_{G(1,2)}(G(2),\rho)$,

(3.28.b) $(P_{23};\pi_2,\pi_3)$ is a fibered product $(G(2),\rho)\times_{G(2,3)}(G(3),\rho)$.

Suppose $L \in \mathfrak{C}$, $\pi_{12} : L \longrightarrow P_{12}$ and $\pi_{23} : L \longrightarrow P_{23}$ so that $\pi_2 \circ \pi_{12} = \pi_2 \circ \pi_{23}$. Then the following conditions are equivalent:

(3.29.a) $(L;\pi_{12},\pi_{21})$ is a fibered product $(P_{12},\pi_2)\times_{G(2)}(P_{23},\pi_2)$,

(3.29.b) $(L;\pi_{12},\pi_3\circ\pi_{23})$ is a fibered product $(P_{12},\rho\circ\pi_2)\times_{G(2,3)}(G(3),\rho)$,

(3.29.c) $(L;\pi_1\circ\pi_{12},\pi_{23})$ is a fibered product $(G(1),\rho)\times_{G(1,2)}(P_{23},\rho\circ\pi_2)$,

(3.29.d) $(L;\pi_1\circ\pi_{12},\pi_2\circ\pi_{12},\pi_3\circ\pi_{23})$ is an inverse limit of G.

In particular, if $G(2,\rho,(1,2))$ and $G(3,\rho,(2,3))$ are pullback bases, or if $G(1,\rho,(1,2))$ and $G(2,\rho,(2,3))$ are pullback bases, then an inverse limit of G exists.

Proof: Proposition 1.10 implies equivalences of (3.29.a,b,c).

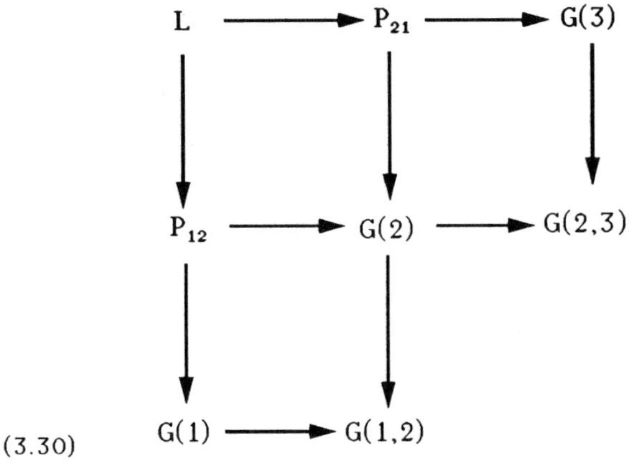

(3.30)

Equivalence of (3.29.a) and (3.29.d) is a routine diagram chase. □

Corollary 3.31: Let $\mathcal{D}$ be a category and let G be a W-graph of $\mathcal{D}$-objects. Suppose $A\in\mathcal{D}$, $\alpha_{12}:G(1,2)\longrightarrow A$ and $\alpha_{23}:G(2,3)\longrightarrow A$ are morphisms such that $(G(2);\rho_{12},\rho_{23})$ is a fibered product for $(G(1,2),\alpha_{12})\times_A(G(2,3),\alpha_{23})$. If either

(3.32.a) α_{12} and $G(1,\rho,(1,2))$ are pullback bases, or

(3.32.b) α_{23} and $G(3,\rho,(2,3))$ are pullback bases,

then an inverse limit of G exists. Moreover, if $(L;\pi_1,\pi_2,\pi_3)$ is an inverse limit of G, then $(L;\pi_1,\pi_3)$ is a fibered product for $(G(1),\alpha_{12}\circ\rho)\times_A(G(3),\alpha_{23}\circ\rho)$.

Proof: The situation reprises diagram (1.16). □

Proposition 3.33: Let $F\in$ Can.

(A) Let $i,j \in \Lambda(F)$. The $(F(i,j)^P;\rho_i{}^P,\rho_j{}^P)$ is a fibered product $(F(i)^P,\iota_i)\times_F(F(j)^P,\iota_j)$.

(B) Let $A \in \mathcal{C}$, $j \in \Lambda(F)$ and $f \in \mathrm{Mor}_{\mathcal{C}}(A,F(j))$. Suppose f is a pullback base of $\mathcal{C}$. Let $B \in \mathcal{C}$ and let $b \in \mathrm{Mor}_{\mathrm{Can}}(B^P,F)$. Then there is a pullback of $[f,j]$ along b, and the underlying object of any choice of pullback is affine.

Proof: Corollary 3.24 reduces (A) to a tautology about the equivalence relation on chart maps. Once (A) is established, (B) follows from Corollary 3.25 and Corollary 3.31. □

Theorem 3.34: Fix $A \in \mathrm{Can}$ and $(S,\sigma_0),(T,\tau_0) \in \mathrm{Can}/A$. Let $\sigma^0 : s \longmapsto (\sigma(s),i(s))$ represent σ and $\tau^0 : t \longmapsto (\tau(t),j(t))$ represent τ. For $s,s' \in \Lambda(S)$, let $\sigma(s,s') : S(s,s') \longrightarrow A(i(s),i(s'))$ be the unique morphism such that $\rho_{i(s)}\circ\sigma(s,s') = \sigma(s)\circ\rho_s$ and $\rho_{i(s')}\circ\sigma(s,s') = \sigma(s')\circ\rho_{s'}$. For $t,t' \in \Lambda(T)$, let $\tau(t,t') : T(t,t') \longrightarrow A(j(t),j(t'))$ be the unique morpism such that $\rho_{j(t)}\circ\tau(t,t') = \tau(t)\circ\rho_t$ and $\rho_{j(t')}\circ\tau(t,t') = \tau(t')\circ\rho_{t'}$.

Put $\Lambda = \Lambda(S)\times\Lambda(T)$. For $(s,t) \in \Lambda$, let $WP_{s,t}$ be the W-graph G given by

(3.35)
$$\begin{aligned} &G(1) = S(s), && G(2) = A(i(s),j(t)),\\ &G(3) = T(t), &&\\ &G(1,2) = A(i(s)), && G(2,3) = A(j(t)),\\ &G(1,\rho,(1,2)) = \sigma(s), && G(2,\rho,(1,2)) = \rho_{i(s)},\\ &G(2,\rho,(2,3)) = \rho_{j(t)}, && G(3,\rho,(2,3)) = \tau(t). \end{aligned}$$

For $(s,t),(s',t') \in \Lambda$, let $WI_{s,s';t,t'}$ be the W-graph H given by

(3.36)
$$\begin{aligned} &H(1) = S(s,s'), && H(2) = A(i(s),j(t)),\\ &H(3) = T(t,t'), &&\\ &H(1,2) = A(i(s)), && H(2,3) = A(j(t)),\\ &H(1,\rho,(1,2)) = \sigma(s)\circ\rho_{s}, && H(2,\rho,(1,2)) = \rho_{i(s)},\\ &H(2,\rho,(2,3)) = \rho_{j(t)}, && H(3,\rho,(2,3)) = \tau(t)\circ\rho_t. \end{aligned}$$

Observe that if $\tau(t)$ is a pullback base for each $t \in \Lambda(T)$, then each of the above graphs admits an inverse limit.

Now suppose $\tau(t)$ is a pullback base for each $t \in \Lambda(T)$. Select

(3.37.a) for each $(s,t) \in \Lambda$, $(P(s,t);\pi_s,\pi_A,\pi_t)$ an inverse limit of $WP_{s,t}$,

(3.37.b) for each $((s,t),(s',t')) \in \Lambda^2$, $(I(s,s';t,t');\pi_S,\pi_A,\pi_T)$ an inverse limit of $WI_{s,s';t,t'}$.

There is a graph P_0 of $\mathcal{C}$-objects of type $\mathrm{Int}(\Lambda)$ determined as follows

(3.38.a) $P_0((s,t)) = P(s,t)$ for $(s,t) \in \Lambda$,

(3.38.b) $P_0((s,t),(s',t')) = I(s,s';t,t')$ for $((s,t),(s',t')) \in \Lambda^2$,

(3.38.c) for $((s,t),(s',t')) \in \Lambda^2$, projection ρ_1 on $P_0((s,t),(s',t'))$ is the unique morphism such that

$$\pi_s \circ \rho_1 = \rho_s \circ \pi_S \quad \text{and} \quad \pi_t \circ \rho_1 = \rho_t \circ \pi_T,$$

and ρ_2 is the unique morphism such that

$$\pi_{s'} \circ \rho_2 = \rho_{s'} \circ \pi_S \quad \text{and} \quad \pi_{t'} \circ \rho_2 = \rho_{t'} \circ \pi_T.$$

Then

(3.39.a) P_0 is a canopy,

(3.39.b) the assignments $\pi^S : (s,t) \mapsto [\pi_s, s]$ and $\pi^T : (s,t) \mapsto [\pi_t, t]$ determine Can-morphisms $P_0 \longrightarrow S$ and $P_0 \longrightarrow T$,

(3.39.c) $(P_0; \pi^S, \pi^T)$ is a fibered product for $(S,\sigma_0) \times_A (T,\tau_0)$.

Proof: Assume $\tau(t)$ is a pullback base for each $t \in \Lambda(T)$, and fix inverse limits as indicated. We freely identify $s \in \Lambda(S)$ with $i(s) \in \Lambda(A)$; for example, writing $A(s)$ for $A(i(s))$ and ρ_s for $\rho_{i(s)}$. Meaning should always be clear from context. The function j is also suppressed.

Condition (3.5.b) on the canopy A implies that for $X \in \mathcal{C}$, $(s,t) \in \Lambda$ and $f,g \in \mathrm{Mor}_{\mathcal{C}}(X,P(s,t))$,

(3.40) $\quad (\pi_s \circ f = \pi_s \circ g \text{ and } \pi_t \circ f = \pi_t \circ g) \Rightarrow f = g.$

Consequently, if $X \in \mathcal{C}$, $(s,t) \in \Lambda$, $\alpha \in \mathrm{Mor}_{\mathcal{C}}(X,S(s))$ and $\beta \in \mathrm{Mor}_{\mathcal{C}}(X,T(t))$ such that $(\sigma(s) \circ \alpha, s) \sim (\tau(t) \circ \beta, t)$, then there is a unique $\gamma \in \mathrm{Mor}_{\mathcal{C}}(X,P(s,t))$ for which

(3.41) $\quad \pi_s \circ \gamma = \alpha \text{ and } \pi_t \circ \gamma = \beta$

which we denote by $\alpha * \beta$. The analoguous remark holds for morphisms into $I(s,s';t,t')$ with respect to indices $(s,t),(s',t') \in \Lambda$.

Suppose $(s,t),(s',t') \in \Lambda$. On $I(s,s';t,t')$,

(3.42) $\quad \rho_s \circ \pi_A = \sigma(s) \circ \rho_s \circ \pi_S = \rho_s \circ \sigma(s,s') \circ \pi_S$ and

$\quad \rho_t \circ \pi_A = \tau(t) \circ \rho_t \circ \pi_T = \rho_t \circ \tau(t,t') \circ \pi_T$,

$\Rightarrow \quad (\rho_s \circ \sigma(s,s') \circ \pi_S, s)$, $(\rho_{s'} \circ \sigma(s,s') \circ \pi_S, t)$, $(\rho_t \circ \tau(t,t') \circ \pi_T, t)$ and $(\rho_{t'} \circ \tau(t,t') \circ \pi_T, t')$ are equivalent charts from $I(s,s';t,t')$ to A.

Thus,

(3.43) $\quad \rho_1 = (\rho_s \circ \pi_S)*(\rho_t \circ \pi_T)$ and $\rho_2 = (\rho_{s'} \circ \pi_S)*(\rho_{t'} \circ \pi_T)$

are the unique morphisms which can satisfy (3.39.c). With this definition, Proposition

3.27 and Proposition 2.5 quickly imply that ρ_1 is a formal subset.

Having produced ρ_1 and ρ_2, we can exploit condition (3.5.b) for S and for T. It follows easily that a morphism f into I(s,s';t,t') is uniquely determined by $\rho_1 \circ f$ and $\rho_2 \circ f$.

Let s,s',t,t' be as before, and let $\varphi_S : S(s,s') \longrightarrow S(s',s)$ and $\varphi_T : T(t,t') \longrightarrow T(t',t)$ be the morphisms such that

(3.44) $\rho_s \circ \varphi_S = \rho_{s'}$, $\rho_{s'} \circ \varphi_S = \rho_{s'}$, $\rho_t \circ \varphi_T = \rho_t$ and $\rho_{t'} \circ \varphi_T = \rho_{t'}$.

Then

(3.45) $\varphi = (\varphi_S \circ \pi_S)^*(\varphi_T \circ \pi_T) \in \mathrm{Mor}_{\mathcal{C}}(I(s,s';t,t'), I(s',s;t',t))$

has inverse $(\varphi_S^{-1} \circ \pi_S)^*(\varphi_T^{-1} \circ \pi_T)$. Moreover, on I(s,s';t,t'),

(3.46) $\pi_{s'} \circ \rho_1 \circ \varphi = \rho_{s'} \circ \pi_S \circ \varphi = \rho_{s'} \circ \varphi_S \circ \pi_S = \rho_{s'} \circ \pi_S = \pi_{s'} \circ \rho_2$.

Similar reasoning yields

(3.47) $\pi_s \circ \rho_2 \circ \varphi = \pi_s \circ \rho_1$, $\pi_{t'} \circ \rho_1 \circ \varphi = \pi_{t'} \circ \rho_2$ and $\pi_t \circ \rho_2 \circ \varphi = \pi_t \circ \rho_1$,

$\Rightarrow$ $\rho_1 \circ \varphi = \rho_2$ and $\rho_2 \circ \varphi = \rho_1$.

At this point, ρ_2 on I(s,s';t,t') can be expressed as a composition of an isomorphism and a ρ_1 projection on I(s',s;t,t'). Thus, ρ_2 must be a formal subset.

Let $(s,t) \in \Lambda$. Let $\delta_s : S(s) \longrightarrow S(s,s)$ and $\delta_t : T(t) \longrightarrow T(t,t)$ be diagonal morphisms for their respective canopies. Since

(3.48) $\rho_s \circ \sigma(s) \circ \rho_s \circ \delta_s = \rho_s \circ \sigma(s)$ and $\rho_t \circ \tau(t) \circ \rho_t \circ \delta_t = \rho_t \circ \tau(t)$,

we may define

(3.49) $\delta = (\delta_s \circ \pi_s)^*(\delta_t \circ \pi_t) \in \mathrm{Mor}_{\mathcal{C}}(P(s,t), I(s,s;t,t))$.

Computation of $\pi_s \circ \rho_j \circ \delta$ and $\pi_t \circ \rho_j \circ \delta$ yields $\rho_j \circ \delta = 1_{P(s,t)}$ for $j \in \{1,2\}$.

To finish proof of (3.39.a), it remains only to show existence of transition functions. This task is lengthy

Existence of Transition Morphisms It remains only to construct transition morphisms. For this argument, fix $s,s',s'' \in \Lambda(S)$ and $t,t',t'' \in \Lambda(T)$. Let

(3.50.a) $(S^{\#};\pi,\pi'')$ be a fibered product $S(s,s') \times_{S(s')} S(s',s'')$,

(3.50.b) $\omega_S : S^{\#} \longrightarrow S(s,s'')$ be a transition morphism,

(3.50.c) $(T^{\#};\pi,\pi'')$ be a fibered product $T(t,t') \times_{T(t')} T(t',t'')$, and

(3.50.d) $\omega_T : T^{\#} \longrightarrow T(t,t'')$ be a transition morphism.

A characterization of $I(s,s';t,t') \times_{P(s',t')} I(s',s'';t',t'')$ is needed.

Consider the W-graph G where

(3.51) $G(1) = S^{\#}$, $G(2) = A(s,t)$,

$$G(3) = T^{\#},$$
$$G(1,2) = A(s), \qquad G(2,3) = A(t),$$
$$G(1,\rho,(1,2)) = \sigma(s)\circ\rho_s\circ\pi, \qquad G(2,\rho,(1,2)) = \rho_{s'},$$
$$G(2,\rho,(2,3)) = \rho_t \qquad G(3,\rho,(2,3)) = \tau(t)\circ\rho_t\circ\pi.$$

The projections of $S^{\#}$ and $T^{\#}$ are formal subsets, so there is

(3.52) $(L;\pi^S,\pi^A,\pi^T)$ an inverse limit of G.

As with previous W-graphs, we conclude

(3.53.a) a morphism f into L is uniquely determined by $\pi^S\circ f$ and $\pi^T\circ f$,

(3.53.b) the chart maps from L into A $(\sigma(s)\circ\rho_s\circ\pi\circ\pi^S,s)$, $(\sigma(s')\circ\rho_{s'}\circ\pi\circ\pi^S,s')$, $(\sigma(s")\circ\rho_{s"}\circ\pi"\circ\pi^S,s")$, $(\tau(t)\circ\rho_t\circ\pi\circ\pi^T,t)$, $(\tau(t')\circ\rho_{t'}\circ\pi\circ\pi^T,t')$ and $(\tau(t")\circ\rho_t\circ\pi"\circ\pi^T,t")$ are all equivalent.

Consequently, there are unique morphisms

(3.54) $\gamma : L\longrightarrow I(s,s';t,t')$, $\gamma' : L\longrightarrow I(s',s";t',t")$ and $\omega : L\longrightarrow I(s,s";t,t")$

such that

(3.55)
$$\pi_S\circ\gamma = \pi\circ\pi^S \quad\text{and}\quad \pi_T\circ\gamma = \pi"\circ\pi^T,$$
$$\pi_S\circ\gamma" = \pi"\circ\pi^S \quad\text{and}\quad \pi_T\circ\gamma" = \pi"\circ\pi^T,$$
$$\pi_S\circ\omega = \omega_S\circ\pi^S \quad\text{and}\quad \pi_T\circ\omega = \omega_T\circ\pi^T.$$

It suffices to show that

(3.56.a) $(L;\gamma,\gamma")$ is a fibered product $I(s,s';t,t')\times_{P(s',t')}I(s',s";t',t")$,

(3.56.b) $(L;\omega,\gamma)$ is a fibered product $I(s,s";t,t")\times_{P(s,t)}I(s,s';t,t')$,

(3.56.c) $(L;\omega,\gamma")$ is a fibered product $I(s,s";t,t")\times_{P(s",t")}I(s',s";t',t")$.

We shall check (3.56.a,b). Verification of (3.56.c) is similar to that for (3.56.b).

Suppose $Z\in\mathcal{C}$, $f\in\mathrm{Mor}_{\mathcal{C}}(Z,I(s,s';t,t'))$ and $g\in\mathrm{Mor}_{\mathcal{C}}(Z,I(s',s";t',t"))$ so that $\rho_{(s',t')}\circ f = \rho_{(s',t')}\circ g$. Now

(3.57) $\rho_{s'}\circ\pi_S\circ f = \pi_{s'}\circ\rho_{(s',t')}\circ f = \pi_{s'}\circ\rho_{(s',t')}\circ g = \rho_{s'}\circ\pi_S\circ g$,

and $\rho_{t'}\circ\pi_T\circ f = \rho_{t'}\circ\pi_T\circ g$ for similar reasons. Define

(3.58)
$$h_S = (\pi_S\circ f)\wedge(\pi_S\circ g)\in\mathrm{Mor}_{\mathcal{C}}(Z,S^{\#}),$$
$$h_T = (\pi_T\circ f)\wedge(\pi_T\circ g)\in\mathrm{Mor}_{\mathcal{C}}(Z,T^{\#}).$$

Now

(3.59) $\sigma(s)\circ\rho_s\circ\pi\circ h_S = \sigma(s)\circ\rho_s\circ\pi_S\circ f = \rho_s\circ\pi_A\circ f$,

$$\tau(t)\circ\rho_t\circ\pi\circ h_T=\tau(t)\circ\rho_t\circ\pi_T\circ f=\rho_t\circ\pi_A\circ f.$$

Thus, there exists a unique $h\in Mor_{\mathcal{C}}(Z,L)$ such that $\pi^S\circ h=h_S$ and $\pi^T\circ h=h_T$. Computing,

(3.60) $\pi_S\circ\gamma\circ h=\pi\circ\pi^S\circ h=\pi\circ h_S=\pi_S\circ f.$

In the same way, one verifies that $\pi_T\circ\gamma\circ h=\pi_T\circ f$, $\pi_S\circ\gamma''\circ h=\pi_S\circ g$ and $\pi_T\circ\gamma''\circ h=\pi_T\circ g$. Hence, $\gamma\circ h=f$ and $\gamma''\circ h=g$. Next, suppose $H\in Mor_{\mathcal{C}}(Z,L)$ so that $\gamma\circ H=f$ and $\gamma''\circ H=g$. In this case,

(3.61) $\pi\circ\pi^S\circ H=\pi_S\circ\gamma\circ H=\pi_S\circ f=\pi\circ h_S=\pi\circ\pi^S\circ h.$

In the same manner, it follows that $\pi''\circ\pi^S\circ H=\pi''\circ\pi^S\circ h$, $\pi\circ\pi^T\circ H=\pi\circ\pi^T\circ h$ and $\pi''\circ\pi^T\circ H=\pi''\circ\pi^T\circ h$. We deduce that $\pi^S\circ H=\pi^S\circ h$ and $\pi^T\circ H=\pi^T\circ h$. From this, we determine $H=h$. Claim (3.56.a) follows.

Finally, let $Z\in\mathcal{C}$, $f\in Mor_{\mathcal{C}}(Z,I(s,s'';t,t''))$ and $g\in Mor_{\mathcal{C}}(Z,I(s,s';t,t'))$ so that $\rho_{(s,t)}\circ f=\rho_{(s,t)}\circ g$. Now

(3.62) $\rho_s\circ\pi_S\circ f=\pi_s\circ\rho_{(s,t)}\circ f=\pi_s\circ\rho_{(s,t)}\circ g=\rho_s\circ\pi_S\circ g,$

and $\rho_t\circ\pi_T\circ f=\rho_t\circ\pi_T\circ g$ for similar reasons. Define $h_S\in Mor_{\mathcal{C}}(Z,S^{\#})$ and $h_T\in Mor_{\mathcal{C}}(Z,T^{\#})$ by the properties

(3.63) $\omega_S\circ h_S=\pi_S\circ f$ and $\pi\circ h_S=\pi_S\circ g,$

$\omega_T\circ h_T=\pi_T\circ f$ and $\pi\circ h_T=\pi_T\circ g.$

Now

(3.64) $\sigma(s)\circ\rho_S\circ\pi\circ h_S=\sigma(s)\circ\rho_S\circ\pi_S\circ g=\rho_S\circ\pi_A\circ g,$

$\tau(t)\circ\rho_t\circ\pi\circ h_T=\tau(t)\circ\rho_t\circ\pi_T\circ g=\rho_t\circ\pi_A\circ g.$

Thus, there exists a unique $h\in Mor_{\mathcal{C}}(Z,L)$ such that $\pi^S\circ h=h_S$ and $\pi^T\circ h=h_T$. Computing,

(3.65) $\pi_S\circ\gamma\circ h=\pi\circ\pi^S\circ h=\pi\circ h_S=\pi_S\circ g.$

In the same way, one verifies that $\pi_T\circ\gamma\circ h=\pi_T\circ g$; hence $\gamma\circ h=g$.

(3.66) $\pi_S\circ\omega\circ h=\omega_S\circ\pi^S\circ h=\omega_S\circ h_S=\pi_S\circ f.$

Reasoning as above, we deduce that $\omega\circ h=f$. Next, suppose $H\subset Mor_{\mathcal{C}}(Z,L)$ so that $\gamma\circ H=g$ and $\omega\circ H=f$. In this case,

(3.67) $\pi\circ\pi^S\circ H=\pi_S\circ\gamma\circ H=\pi_S\circ g=\pi\circ h_S=\pi\circ\pi^S\circ h.$

Exploiting (3.63) and (3.55) in the same way, we get $\omega_S\circ\pi^S\circ H=\omega_S\circ\pi^S\circ h$, $\pi\circ\pi^T\circ H=\pi\circ\pi^T\circ h$ and $\omega_T\circ\pi^T\circ H=\omega_T\circ\pi^T\circ h$. We deduce that $\pi^S\circ H=\pi^S\circ h$ and $\pi^T\circ H=\pi^T\circ h$. Hence $H=h$. Claim (3.56.b) follows.

Proof of (3.39.b) Hereafter, π^S and π^T are interpreted as in statement (3.39.b). Let $(s,t),(s',t') \in \Lambda$. On $I(s,s';t,t')$,

$$\text{(3.68)} \qquad \pi_s \circ \rho_{(s,t)} = \rho_s \circ \pi_S \quad \text{and} \quad \pi_{s'} \circ \rho_{(s',t')} = \rho_{s'} \circ \pi_S$$

$$\Rightarrow \qquad (\pi_s \circ \rho_{(s,t)}, s) \text{ and } (\pi_{s'} \circ \rho_{(s',t')}, s') \text{ are equivalent.}$$

A similar tautology justifies definition of π^T.

Proof of (3.39.c) Let $(s,t) \in \Lambda$.

$$\text{(3.69)} \qquad (\sigma^0 \circ \pi^S)(s,t) = [\sigma(s) \circ \pi_s, s] = [\rho_s \circ \pi_A, s] = [\rho_t \circ \pi_A, t] = [\tau(t) \circ \pi_t, t]$$
$$= (\tau^0 \circ \pi^T)(s,t).$$

It follows that $\sigma^0 \circ \pi^S = \tau^0 \circ \pi^T$.

To check that P_0 is a fibered product, it suffices to prove that every source from an object B^p for $B \in \mathfrak{C}$ factors uniquely through P_0. For the rest of the argument, fix $B \in \mathfrak{C}$, $\alpha \in Cc(B,S)$, $\beta \in Cc(B,T)$, $(f,s) \in \alpha$ and $(g,t) \in \beta$ such that $\sigma^0 \circ \alpha = \tau^0 \circ \beta$. Then $(\sigma(s) \circ f, i(s))$ and $(\tau(t) \circ g, j(t))$ are equivalent chart maps. Consequently, there is a unique $h \in Mor_{\mathfrak{C}}(B,P(s,t))$ such that $\pi_s \circ h = f$ and $\pi_t \circ h = g$. By inspection, $h_0 = [h,(s,t)] \in Mor_{Can}(B^p,P_0)$ satisfies $\pi^S \circ h_0 = \alpha$ and $\pi^T \circ h_0 = \beta$.

Suppose $H_0 \in Mor_{Can}(B^p,P_0)$ so $\pi^S \circ H_0 = \alpha$ and $\pi^T \circ H_0 = \beta$. Express H_0 as $[H,(s',t')] \in Cc(B,P_0)$. Now $[\pi_{s'} \circ H, s'] = [\pi_s \circ h, s]$, so there is $k_S \in Mor_{\mathfrak{C}}(B,S(s,s'))$ such that $\rho_s \circ k_S = \pi_s \circ h$ and $\rho_{s'} \circ k_S = \pi_{s'} \circ H$. Similarly, there must be k_T such that $\rho_t \circ k_T = \pi_t \circ h$ and $\rho_{t'} \circ k_T = \pi_{t'} \circ H$. Tautologically, there exists $k \in Mor_{\mathfrak{C}}(B,I(s,s';t,t'))$ so that $\pi_S \circ k = k_S$ and $\pi_T \circ k = k_T$. It follows that

$$\text{(3.70)} \qquad \rho_{(s,t)} \circ k = h \quad \text{and} \quad \rho_{(s',t')} \circ k = H$$

$$\Rightarrow \qquad [h,(s,t)] = [H,(s',t')] \quad \Rightarrow \quad h_0 = H_0.$$

We are done. □

<u>Corollary 3.71:</u> (A) The pasting functor takes pullback bases to pullback bases.

(B) If $\mathfrak{C}$ is closed under arbitrary fibered products, then so is Can.

<u>Definition 3.72:</u> Let $b_0 : B_0 \longrightarrow C_0$ be an Can-morphism. We say that b_0 is a <u>formal pullback</u>

base if $b_0(j)$ can be represented by a pullback base of $\mathfrak{C}$ for each $j \in \Lambda(B_0)$. Clearly the composition of two formal pullback bases is also a formal pullback base. In the construction of Theorem 3.34, the morphism π^S is a formal pullback base; however, we cannot deduce that every pullback of a formal pullback base is a formal pullback.

Corollary 3.74: A formal pullback base is a pullback base with respect to Can.

§4 Lifting a Universe of Subsets to Canopies

Throughout this section, assume hypothesis (3.1). Put $\mathrm{Can} = \mathrm{Can}(\mathfrak{C})$.

Let Aux ('auxiliary') be a universe of subsets for $\mathfrak{C}$. The case Aux = Sub and that where Aux is a universe of layered morphisms are of particular interest. Let $b_0 : B_0 \longrightarrow A_0$ be a Can-morphism, $C \in \mathfrak{C}$, and $c_0 \in \mathrm{Mor}_{\mathfrak{C}}(C^P, A_0)$. An Aux-splitting for c_0 along b_0 is a triple (P, π_B, π_C) such that $P \in \mathfrak{C}$, $\pi_C \in \mathrm{Mor}_{\mathfrak{C}}(P, C)$ and $\pi_B \in \mathrm{Mor}_{\mathrm{Can}}(P^P, B_0)$ such that

(4.1.a) $(P^P; \pi_B, \pi_C{}^P)$ is a pullback $b_0^{-1}(C^P, c_0)$,

(4.1.b) $\pi_C \in \mathrm{Aux}(P, C)$.

We say c_0 Aux-splits (along b_0) if such a splitting exists. If c_0 Aux-splits and if (Q, p, q) satisfies the analogue of (4.1.a), then $q \in \mathrm{Aux}(Q, C)$.

If $j \in \Lambda(A_0)$ and ι_j Aux-splits, we also say that j Aux-splits. The morphism b_0 is called Aux-like if each $j \in \Lambda(A_0)$ Aux-splits. The immediate objective is to show that Aux-like morphisms form a universe of subsets for Can.

Proposition 4.2: Let Aux be a universe of subsets for $\mathfrak{C}$.

(A) Let b be a $\mathfrak{C}$-morphism. Then b is an Aux-morphism if and only if b^P is Aux-like.

(B) A Can-isomorphism is Aux-like.

(C) The compostion of Aux-like Can-morphisms is Aux-like.

(D) Suppose Aux is a universe of embeddings. If $b_0 : B_0 \longrightarrow C_0$ and $c_0 : C_0 \longrightarrow D_0$ are Can-morphisms such that c_0 and $c_0 \circ b_0$ are Aux-like, then b_0 is Aux-like.

Proof: (A) and (B) are tautological.

(C) Let $b_0 : B_0 \longrightarrow C_0$ and $c_0 : C_0 \longrightarrow D_0$ be Aux-like Can-morphisms, and let $k \in \Lambda(D_0)$.

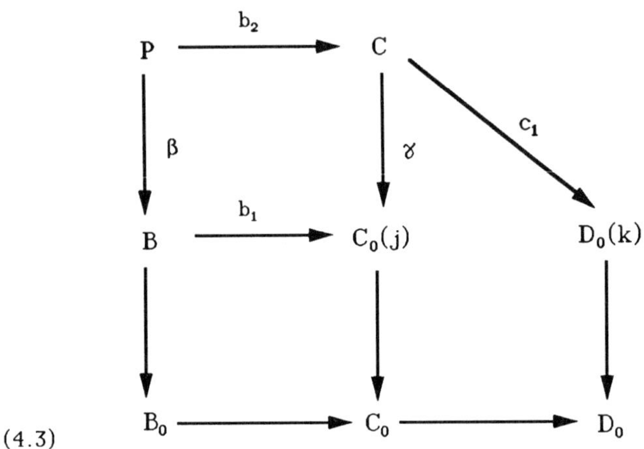

(4.3)

Let (C,π,c_1) be an Aux-splitting at k for c_0. Represent π by (γ,j), and let (B,τ,b_1) be an Aux-splitting at j. As $b_1 \in$ Aux, there is a fibered product $(P;\beta,b_2)$ for $(B,b_1)\times_{C(j)}(C,\gamma)$ in which $b_2 \in$ Aux. Proposition 1.10 and Corollary 3.25 imply that $(P,\tau\circ\beta^P,c_1\circ b_2)$ is an Aux-splitting at k for $c_0\circ b_0$.

(D) Suppose Aux has the added property (2.4). Let $b_0 : B_0 \longrightarrow C_0$ and $c_0 : C_0 \longrightarrow D_0$ be Can-morphisms such that $c_0\circ b_0$ and c_0 are Aux-like. Let $j \in \Lambda(C_0)$, and let (c,k) represent $c_0(j)$.

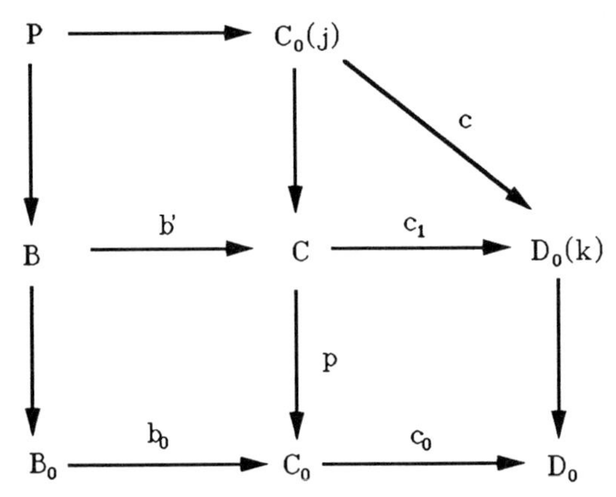

(4.4)

Let $(C;\pi_C,c_1)$ and (B,π_B,b_1) be Aux-splittings of k along c_0 and $c_0\circ b_0$, respectively. Let $b' \in \mathrm{Mor}_{\mathfrak{C}}(B,C)$ such that $b'^P = b_0\times 1_{D_0(k)^P}$. As $c_1\circ b' = b_1$ and $c_1,b_1 \in$ Aux, $b' \in$ Aux as well. Put $\gamma \in \mathrm{Mor}_{\mathfrak{C}}(C_0(j),C)$ so $\gamma^P = \iota_j \wedge c^P$. Let $(P;\sigma_B,\pi_C)$ be a fibered product $(B,b')\times_C(C_0(j),\gamma)$, and then $\pi_C \in$ Aux. Finally, Proposition 1.10 assures that $(P;\pi_B\circ\sigma_B{}^P,\gamma)$ is an Aux-splitting of b_0 along j. □

Corollary 4.5: (A) For $B_0 \in$ Can and $j \in \Lambda(B_0)$, ι_j is Sub-like.

(B) Suppose Sub is a universe of embeddings. If Θ is a half-map which represents and Sub-like Can-morphism $b_0 : B_0 \longrightarrow C_0$, then for each $j \in \Lambda(F)$, $\Theta(j) \in$ Sub.

Proof: (A) is a consequence of Corollary 3.33.

(B) Let Θ and b_0 be as hypotheiszed. Let $j \in \Lambda(B_0)$, and express $\Theta(j) = (g,k)$. Then $b_0 \circ \iota_b = [g,k] = g^p \circ \iota_k$. From Proposition 4.2.D, we conclude that g^p must be Sub-like. Proposition 4.2.A finishes the proof. □

Theorem 3.34 offers an excellent existence criterion for fibered products, but its characterization is cumbersome. The next definition is part of a more useful viewpoint.

Definition 4.6: Let $C_0 \in$ Can. By a pullback system into C_0, we mean a pair (Q_0,q) such that

(4.7.a) $Q_0 \in \mathcal{C}^p$ and $\Lambda(Q_0) = \Lambda(C_0)$,

(4.7.b) q assigns to each $j \in \Lambda(C_0)$ a morphism $q(j) \in \mathrm{Mor}_{\mathcal{C}}(Q_0(j),C_0(j))$,

(4.7.c) for $j,k \in \Lambda(C_0)$, there is $q' \in \mathrm{Mor}_{\mathcal{C}}(Q_0(j,k),C_0(j,k))$ such that

$(Q_0(j,k);\rho_j,q')$ is a fibered product for

$$(Q_0(j),q(j)) \times_{C_0(j)} (C_0(j,k),\rho_j), \text{ and}$$

$(Q_0(j,k);q',\rho_k)$ is a fibered product

$$(C_0(j,k),\rho_k) \times_{C_0(k)} (Q_0(k),q(k)).$$

Obviously a morphism q' of the type described in (4.7.c) is uniquely determined by $j,k \in \Lambda(C_0)$.

Most pullback systems arise from pulling back along morphisms. Suppose $A_0 \in$ Can and $(B_0,b_0),(C_0,c_0) \in \mathrm{Can}/A_0$, and that for each $j \in \Lambda(C_0)$ there are

(4.8) $Q(j) \in \mathcal{C}$, $q(j) \in \mathrm{Mor}_{\mathcal{C}}(Q(j),C_0(j))$, and

$\pi(j) \in \mathrm{Mor}_{\mathrm{Can}}(Q(j)^p,B_0)$.

such that

(4.9) $(Q(j)^P;\pi(j)^P,q(j))$ is a pullback $b_0^{-1}(C_0(j),c_0\circ\iota_j)$.

By Proposition 1.10, for $j,k \in \Lambda(C_0)$, there is

(4.10) $Q(j,k) \in \mathcal{C}$, $q(j,k) \in \mathrm{Mor}_{\mathcal{C}}(Q(j,k),C_0(j,k))$, $\pi(j,k) \in \mathrm{Mor}_{\mathrm{Can}}(Q(j)^P,B_0)$.

such that

(4.11) $(Q(j,k)^P;\pi(j,k)^P,q(j,k))$ is a pullback $b_0^{-1}(C_0(j,k),c_0\circ\iota_j\circ\rho_j{}^P)$.

This function Q which has been constructed on $\Lambda(C_0)$ extends to an intersection graph Q_0 by assigning morphisms determined by

(4.12) $\forall j,k \in \Lambda(C_0)$,

$Q_0[j,k](\rho_1)^P$ is the appropriate $b_0^{-1}\{C_0[j,k](\rho_1)\}$, and

$Q_0[j,k](\rho_2)^P$ is the appropriate $b_0^{-1}\{C_0[j,k](\rho_2)\}$.

Trivially, $(Q_0,j\mapsto q(j))$ is a pullback system. We refer to this as a <u>pullback system of c_0 along b_0</u>. Tautologically, the function $j\mapsto\pi(j)$ is an Can-morphism, which we call the canonical projection.

<u>Theorem 4.13:</u> Let $A_0 \in \mathrm{Can}$, and let $(B_0,b_0),(C_0,c_0) \in \mathrm{Can}/A_0$. Suppose (Q_0,q) is a pullback system of c_0 along b_0, and let $\pi^\#$ denote the canonical morphism $Q_0 \longrightarrow B_0$. Let $q^\#$ denote the function $j\mapsto[q(j),j]$ on $\Lambda(Q_0)$. Then $q^\# \in \mathrm{Mor}_{\mathrm{Can}}(Q_0,C_0)$, and $(Q_0;\pi^\#,q^\#)$ is a fibered product $(B_0,b_0)\times_{A_0}(C_0,c_0)$. In particular, if $c_0 = 1_{A_0}$, then $\pi^\#$ is an isomorphism.

<u>Proof:</u> Tautologically, $c_0\circ q^\# = b_0\circ\pi^\#$. For $j,k \in \Lambda(C_0)$, let $q(j,k): Q_0(j,k) \longrightarrow C_0(j,k)$ be the unique morphism such that $\rho_j\circ q(j,k) = q(j)\circ\rho_j$ and $\rho_k\circ q(j,k) = q(k)\circ\rho_k$, and let $\pi(j,k) = \pi(j)\circ\rho_j{}^P = \pi_k\circ\rho_k{}^P$ on $Q_0(j,k)$.

Suppose $D \in \mathcal{C}$ and $f_0,g_0 \in \mathrm{Mor}_{\mathrm{Can}}(D^P,Q_0)$ such that $\pi^\#\circ f_0 = \pi^\#\circ g_0$ and $q^\#\circ f_0 = q^\#\circ g_0$. Fix representatives (f,j) and (g,k) for f_0 and g_0. The latter condition implies existence of $h_1 \in \mathrm{Mor}_{\mathcal{C}}(D,C_0(j,k))$ so $\rho_j\circ h_1 = q(j)\circ f$ and $\rho_k\circ h_1 = q(k)\circ g$. Let $h \in \mathrm{Mor}_{\mathcal{C}}(D,Q_0(j,k))$ be the unique morphism such that

(4.14) $q(j,k)\circ h = h_1$ and $\pi^\#(j)\circ\rho_j{}^P\circ h^P = \pi^\#(j)\circ f^P = \pi^\#(k)\circ g^P$.

Obviously,

(4.15) $\pi^\#(j)\circ f^P = \pi^\#(j)\circ(\rho_j\circ h)^P$ and

$q(j)\circ f = \rho_j\circ q(j,k)\circ h = q(j)\circ\rho_j\circ h$,

$\pi^\#(k)\circ g^P = \pi^\#(k)\circ(\rho_k\circ h)^P$ and $q(k)\circ g = q(j)\circ\rho_k\circ h$.

As Q(j) and Q(k) are derived as pullbacks, $\rho_j\circ h = f$ and $\rho_k\circ h = g$. Thus, $f_0 = g_0$.

Suppose $D \in \mathfrak{C}$, and $f_0 : D^p \longrightarrow B_0$ and $g_0 : D^p \longrightarrow C_0$ such that $b_0 \circ f_0 = c_0 \circ g_0$. Fix a representative (g,k) of g_0, respectively. Then we may take $h : D \longrightarrow Q(k)$ such that

(4.16) $\quad h^p = f_0 \wedge g^p \in \mathrm{Mor}_{\mathrm{Can}}(D^p, Q(k))$.

Tautologically, $\pi^{\#} \circ [h,k] = f_0$ and $q^{\#} \circ [h,k] = [g,k] = g_0$. □

Corollary 4.17: Suppose Aux is a universe of subsets for $\mathfrak{C}$. Let Aux_0 denote the class of Aux-like morphisms.

(A) Aux_0 is a universe of subsets for Can. If Aux is a universe of embeddings for $\mathfrak{C}$, then Aux_0 is a universe of embeddings for Can. We refer to Aux_0 as the lift of Aux.

(B) If Sub and Aux satisfy the smallness condition, then Aux_0 satisfies the smallness condition.

Proof: Suppose $A_0 \in \mathrm{Can}$. Theorem 4.13 states that each Aux-like morphism $b_0 : B_0 \longrightarrow A_0$ is Can/A_0-isomorphic to an Aux-like morphism $b_1 : B_1 \longrightarrow A_0$ such that $\Lambda(B_1) = \Lambda(A_0)$ and b_1 can be represented by a half-map whose values are in Aux. The latter b_1 is a formal pullback base! This normalized form of b_0 implies that each Aux-like morphism is a pullback base. Also, once (A) is established, this reduction easily implies (B).

Suppose $b_0 : B_0 \longrightarrow A_0$ is Aux-like and $c_0 : C_0 \longrightarrow A_0$ is any morphism. Let $(P;\pi_C,\pi_B)$ be any pullback $c_0^{-1}b_0$. To show that π_C is Aux-like, it is necessary to consider pullbacks of it along affine morphisms into C_0. But, such pullbacks are also pullbacks of b_0 along affine morphisms into B_0. An affine morphism into B_0 is a composition of a $\mathfrak{C}$-morphism (under pasting) with a coordinate chart, and it is assumed that b_0 Aux-splits along every chart. It follows that b_0 splits along every affine morphism. This proves that π_C is also Aux-like. Proposition 4.2 finishes the proof that Aux_0 is a universe of subsets. □

Corollary 4.18: Let Aux be a universe of subsets for $\mathfrak{C}$, and let $b_0 : B_0 \longrightarrow C_0$ be an Aux-like morphism. If C_0 is affine, then B_0 is affine.

§5 A Topology for Canopies

We continue under (3.1).

Put Can = Can($\mathcal{C}$). Let Sub^p denote the lift of Sub to Can. Suppose $B_0 \in$ Can and Θ is an indexed subset of Can/B_0 such that $\Theta(x) \in \text{Sub}^p$ for each $x \in \text{dom}(\Theta)$. For $j \in \Lambda(B_0)$, a splitting of Θ along j is any function φ which to each $x \in \text{dom}(\Theta)$ assigns a Sub-splitting of $\Theta(x)$ along ι_j. A splitting of a family exists along each $j \in \Lambda(B_0)$; any two are isomorphic as families of $\mathcal{C}/B_0(j)$. We say Θ covers along j if there is a splitting $x \mapsto (Q(x), \pi_x, q_x)$ such that $x \mapsto (Q(x), q_x)$ is a cover of $B_0(j)$ with respect to Cov. For the moment, we say Θ is cover-like if Θ covers along each $j \in \Lambda(B_0)$.

Lemma 5.1: Let $B_0 \in$ Can. Then the family $j \mapsto (B_0(j)^p, \iota_j, B_0)$ on $\Lambda(B_0)$ is cover-like. We refer to this as the assigned cover of B_0.

Proof: Let Θ be the indicated family. For $k \in \Lambda(B_0)$, $(B_0(j,k)^p, \rho_k{}^p)$ is known to be an example of $\iota_k{}^{-1}(B_0(j)^p, \iota_j)$. When $j = k$, ρ_k is a covering morphism. □

Theorem 5.2: Let Cov^p be the class of subsets of Mor(Can) which are commonly supported and cover-like. Then Cov^p is a Grothendieck topology for Can.

(A) If Cov meets the smallness conditions, then so does Cov^p.

(B) If Θ is a non-empty cone in Mor($\mathcal{C}$), then the function $x \mapsto \Theta(x)^p$ is an indexed cover with respect to Cov^p if and only if Θ is an indexed cover with respect to Cov.

(C) Let Cvm = $\text{Cvm}_{\mathcal{C}}$. Then a Can-morphism is a covering morphism with respect to Cov^p if and only if it is Cvm-like.

(D) If Cov is flush, then so is Cov^p.

Remark 5.3: Once the theorem is established, we cease using the term 'cover-like', and regard Can as topologized in this manner. The triple of data (Can, Sub^p, Cov^p) is denoted by $\mathcal{C}^p$, and called the pasting lift of the original data.

Proof: The essential point is to prove that Cov^p is a Grothendieck topology.

Conditions (2.9.a,b,c) for $\mathrm{Cov}^{\mathrm{p}}$ are tautological.

Suppose Θ is a cover-like cone in Can/B_0 for $B_0 \in \mathrm{Can}$, and suppose $A \in \mathcal{C}$ and $b_0 \in \mathrm{Mor}_{\mathrm{Can}}(A^{\mathrm{p}},B_0)$. There is $j \in \Lambda(B_0)$ and $b : A \longrightarrow B(j)$ so that $b_0 = \iota_j \circ b^{\mathrm{p}}$. Proposition 1.10 implies that any function of the form $x \longmapsto b^{-1}\Theta(x)$ on $\mathrm{dom}(\Theta)$ is the image of an indexed cover of A under pasting. This observation easily implies Conditions (2.9.d,e).

Suppose $b_0 : B_0 \longrightarrow A_0$ is Sub-like and $s : A_0 \longrightarrow B_0$ such that $b_0 \circ s = 1_{A_0}$. To complete the proof that $\mathrm{Cov}^{\mathrm{p}}$ is topology, it suffices to show that $\iota_j^{-1} b_0$ splits to a $\mathrm{Cvm}_{\mathcal{C}}$-morphism for each $j \in \Lambda(A_0)$. For $j \in \Lambda(B_0)$, $\iota_j^{-1}(b \circ s) \approx \iota_j$, and the conclusion follows by pulling back s to a $\mathcal{C}$-morphism.

Theorem 4.17 quickly yields (A). Parts (B) and (C) are vacuous, and (D) is a trivial consequence of definition and Proposition 1.10. □

§6 Functorial Properties of Canopies

We continue under (3.1). Obviously the pasting functor can be characterized in terms of a universal property involving canopies. However, $\mathfrak{C}^p$ also has a mapping property with respect to functors of sections.

Proposition 6.1: Assume (3.1) and let $\Gamma : \mathfrak{C} \longrightarrow \mathcal{D}$ be a functor of sections. Let p_0 denote the pasting functor.

(A) If Φ_0 and Φ_1 are functors of sections $\mathfrak{C}^p \longrightarrow \mathcal{D}$ such that $\Phi_0 \circ p_0$ and $\Phi_1 \circ p_0$ are functorially equivalent, then Φ_0 and Φ_1 are functorally equivalent.

(B) Suppose for each canopy G over $\mathfrak{C}$, the graph $\Gamma(G)$ admits a colimit in $\mathcal{D}$. Then there exists a functor of sections $\Phi : \mathfrak{C}^p \longrightarrow \mathcal{D}$ such that $\Phi \circ p_0 = \Gamma$.

Proof: Part (A) is trivial.

(B) Assume that the image each canopy of $\mathfrak{C}$ under Γ admits a colimit. For each $G \in \mathfrak{C}^p$, let $\beta[G] : \Gamma(G) \longrightarrow \Phi_0(G)$ be a colimit; for $C \in \mathfrak{C}$, choose $\Phi(C^p) = \Gamma(C)$ and $\beta[C^p] = 1_{\Gamma(C)}$. For a canopy G, we continue to write $\Gamma(G)$ to signify the image of G regarded as a graph of $\mathcal{D}$-objects.

Suppose $B \in \mathfrak{C}$ and $A_0 \in \mathfrak{C}^p$. Let $(b,j),(c,k) \in Ch(B,A_0)$. If $(b,j) \sim (c,k)$, take $h : B \longrightarrow A_0(j,k)$ so $\rho_j \circ h = b$ and $\rho_k \circ h = c$ and then

$$(6.2) \qquad \beta[A_0](j) \circ \Gamma(b) = \beta[A_0](j) \circ \Gamma(\rho_j) \circ \Gamma(h)$$
$$= \beta[A_0](k) \circ \Gamma(\rho_k) \circ \Gamma(h) = \beta[A_0](k) \circ \Gamma(c).$$

Thus, there is a function from $Cc(B,A_0) \longrightarrow Mor_{\mathcal{D}}(\Gamma(B),\Phi_0(A_0))$ determined by $[b,j] \longmapsto \beta[A_0] \circ \Gamma(b)$. Any function of this type is indicated by 'Φ_1'.

Suppose $b_0 : B_0 \longrightarrow A_0$ is a $\mathfrak{C}^p$-morphism. Then $j \longmapsto \Phi_1(b_0(j))$ on $\Lambda(B_0)$ extends to a cone $\Gamma(B_0) \longrightarrow \Phi_0(A_0)$. Define $\Phi_2(b_0)$ to be the factoring of this cone. Observe that if $B \in \mathfrak{C}$ and $B_0 = B^p$, then $\Phi_2(b_0)$ agrees with $\Phi_1(b')$ where b' is the chart map $b_0(1)$. Regard Φ_0 and Φ_2 as class-theoretic functions on $Obj(\mathfrak{C}^p)$ and $Mor(\mathfrak{C}^p)$, respectively. It is simple to verify that the two determine a covariant functor, which we denote by Φ. Obviously $\Phi \circ p_0 = \Gamma$.

Let $B_0 \in \mathfrak{C}^p$. Suppose Θ is a $\mathfrak{C}^p$-cover of B_0, $D \in \mathcal{D}$ and $f,g \in Mor_{\mathcal{D}}(\Phi(B_0),D)$ so that $f \circ \Phi(\Theta(x)) = g \circ \Phi(\Theta(x))$ for each $x \in dom(\Theta)$. We claim that $f = g$. By inspection, $\Phi(\iota_j) = \beta[B_0](j)$ for each $j \in \Lambda(B_0)$; consequently, it suffices to show that $f \circ \Phi(\iota_j) = g \circ \Phi(\iota_j)$ for each j. Fix $j \in \Lambda(B_0)$, and for each $x \in dom(\Theta)$ let $(C(x),\gamma(x),c(x))$ be a splitting of ι_j

along $\Theta(x)$. Then

(6.3.a) $x \mapsto (C(x),c(x))$ is a $\mathcal{C}$-cover of $B_0(j)$,

(6.3.b) for each $x \in dom(\Theta)$, $f \circ \Phi(i_j) \circ \Gamma(c(x)) = g \circ \Phi(\iota_j) \circ \Gamma(c(x))$.

As Γ is a functor of sections, $f \circ \Phi(\iota_j) = g \circ \Phi(\iota_j)$.

Again, suppose $B_0 \in \mathcal{C}^p$ and Θ is a $\mathcal{C}^p$-cover of B_0. Let $\Theta^{\#} : B_1 \longrightarrow B_0$ be a canopy of Θ. We claim $\Phi(\Theta^{\#})$ is a colimit. Let B_2 denote the image of B_1 under Φ, and suppose $\alpha : B_2 \longrightarrow A$ is a cone in $\mathcal{D}$. For $j \in \Lambda(B_0)$ and $x \in dom(\Theta)$, let $(C(j,x);c(j,x),\gamma(j,x))$ be a fibered product $\iota_j \times_{B_0} \Theta(x)$. For each $j \in \Lambda(B_0)$, let G_j be the image of a canopy of $x \mapsto (C(j,x),c(j,x))$, and let α_j be the unique cone $G_j \longrightarrow A$ such that $\alpha_j(x) = \alpha(x) \circ \gamma(j,x)$ for each $x \in dom(\Theta)$. For $j \in \Lambda(B_0)$, there is a unique $b_j \in Mor_{\mathcal{D}}(\Gamma(B_0(j)),D)$ so that $\alpha_j(x) = b_j \circ \Phi(c(j,x))$ for each $x \in dom(\Theta)$. The preceeding paragraph implies the conclusion once we produce $b : \Phi(B_0) \longrightarrow D$ so that $b_j = b \circ \iota_j$ for each $j \in \Lambda(B_0)$. This reduces the problem to showing that $b_j \circ \Gamma(\rho_j) = b_k \circ \Gamma(\rho_k)$ on $B_0(j,k)$ for $j,k \in \Lambda(B_0)$.

Fix $j,k \in \Lambda(B_0)$. Put $\rho^* = \iota_j \circ \rho_j = \iota_k \circ \rho_k$ on $B_0(j,k)^p$. It suffices to show that for $x,y \in dom(\Theta)$,

(6.4) $b_j \circ \Gamma(\rho_j) \circ \pi = b_k \circ \Gamma(\rho_k) \circ \pi$,

where π is projection $\rho^{*-1}\Theta(x) \times_{B_0(j,k)} \rho^{*-1}\Theta(y)$. However, the π of (6.4) factors through projection $\Theta(x) \times_{B_0} \Theta(y)$. By hypothesis, (6.4) is true with the latter projection in place of π. □

A variant of Proposition 6.1 will be introduced in Section 10.

PART III CANOPIES AND COLIMITS

So far, we have viewed canopies as objects in a larger category. But a $\mathfrak{C}^p$ morphism can be considered as a construct of graphs. If two canopies are $\mathfrak{C}^p$-isomorphic, then a colimit for one natural determines a colimit for the other. In Part III, we study the behavior of canopies as graphs whose limits are of particular interest.

§7 Monomorphisms

We begin with general remarks whose proofs are left to the reader.

Proposition 7.1: Let $\mathfrak{C}$ be a category and let $b: B \longrightarrow C$ be a $\mathfrak{C}$-morphism. Then b is monomorphic if and only if $(B;1_B,1_B)$ is a fibered product $(B,b)\times_C(B,b)$.

Corollary 7.2: Let $\mathfrak{C}$ be a topologized category, and let $b_0 : B_0 \longrightarrow C_0$ be a $\mathfrak{C}^p$-morphism. Then b_0 is monomorphic if and only if for each $A \in \mathfrak{C}$,

(7.3) $\quad f \longmapsto b \circ f \quad$ from $\mathrm{Mor}_{\mathfrak{C}}(A^p,B_0) \longrightarrow \mathrm{Mor}_{\mathfrak{C}}(A^p,C_0)$,

is monomorphic.

Proof: Use Corollary 3.24. □

Corollary 7.4: Let $\mathfrak{C}$ be a topologized category. Then the pasting functor $\mathfrak{C} \longrightarrow \mathfrak{C}^p$ maps $\mathfrak{C}$-monomorphisms to $\mathfrak{C}^p$-monomorphisms.

Proposition 7.5: Let $\mathfrak{C}$ be a category and let $b: B \longrightarrow C$ be a $\mathfrak{C}$-monomorphism. Suppose $X,Y,Z \in \mathfrak{C}$, $x \in \mathrm{Mor}_{\mathfrak{C}}(X,B)$, $y \in \mathrm{Mor}_{\mathfrak{C}}(Y,B)$, $p \in \mathrm{Mor}_{\mathfrak{C}}(Z,X)$ and $q \in \mathrm{Mor}_{\mathfrak{C}}(Z,Y)$. Then $(Z;p,q)$ is a fibered product $(X,x)\times_B(Y,y)$ if and only if $(Z;p,q)$ is a fibered product $(X,b\circ x)\times_C(Y,b\circ y)$.

Proposition 7.6: Let $\mathfrak{C}$ be a category.

(A) A composition of two $\mathfrak{C}$-monomorphisms is monomorphic.

(B) A $\mathfrak{C}$-isomorphism is monomorphic.

(C) Suppose $A,B,C \in \mathfrak{C}$, $a \in \mathrm{Mor}_{\mathfrak{C}}(A,B)$ and $b \in \mathrm{Mor}_{\mathfrak{C}}(B,C)$. If $b\circ a$ is monomorphic, then a is monomorphic.

(D) Suppose $b : B \longrightarrow C$ is a $\mathcal{C}$-monomorphism. Let $d : D \longrightarrow C$ be a $\mathcal{C}$-morphism, and suppose $(P;p,q)$ is a pullback $d^{-1}(B,b)$. Then p is monomorphic.

Proof: Trivial. □

Proposition 7.7: Let $\mathcal{C}$ be a topologized category. Suppose $b_0 : B_0 \longrightarrow A_0$ is a $\mathcal{C}^P$-morphism. Then b_0 is monomorphic if and only if for each $j,k \in \Lambda(B_0)$, $(B_0(j,k)^P;\rho_j{}^P,\rho_k{}^P)$ is a fibered product $(B_0(j)^P, b_0 \circ \iota_j) \times_{A_0} (B_0(k)^P, b_0 \circ \iota_k)$. In particular, the canonical morphism of a canopy of a family of formal subsets is monomorphic.

Proof: Necessity of the condition is established. The converse follows from Corollary 7.6 and the observation that every morphism from an affine object factors through a coordinate chart. □

Suppose $\mathcal{C}$ is a topologized category, $A_0 \in \mathcal{C}^P$ and $A \in \mathcal{C}$. There is a bijection between the set of cones from A_0 (regarded as a graph of $\mathcal{C}$-objects) into A to $\mathrm{Mor}_{\mathcal{C}^P}(A_0, A^P)$ determined by mapping a cone Θ to the function $j \mapsto \Theta(j)^P$ on $\Lambda(A_0)$. We identify these sets in this manner. In particular, we refer to a cone as monomorphic if the corresponding $\mathcal{C}^P$-morphism is monomorphic.

Let us return to two earlier concepts with monomorphisms in mind.

Proposition 7.8: Let $\mathcal{C}$ be a topologized category. Suppose $C_0 \in \mathcal{C}^P$, (Q_0,q) is a pullback system of C_0, and $q(j)$ is a $\mathcal{C}$-monomorphism for each $j \in \Lambda(C_0)$. Then $j \mapsto [q(j),j]$ is $\mathcal{C}^P$-monomorphic.

Proof: We invoke Proposition 7.7. It suffices to show that for

(7.9) $\quad j,k \in \Lambda(C_0), \quad X \in \mathcal{C}, \quad f \in \mathrm{Mor}_{\mathcal{C}}(X,Q_0(j)) \quad \text{and} \quad g \in \mathrm{Mor}_{\mathcal{C}}(X,Q_0(k))$

such that

(7.10) $\quad [q(j) \circ f, j] = [q(k) \circ g, k] \ \text{ in } \ Cc(X,C_0),$

there is $h \in \mathrm{Mor}_{\mathfrak{C}}(X,Q_0(j,k))$ so $\rho_j \circ h = f$ and $\rho_k \circ h = g$.

Assume hypotheses (7.9) and (7.10). Let $\pi : Q_0(j,k) \longrightarrow C_0(j,k)$ be the unique morphism such that $\rho_j \circ \pi = q(j) \circ \rho_j$ and $\rho_k \circ \pi = q(k) \circ \rho_k$. By assumption, there is $\gamma : X \longrightarrow C_0(j,k)$ such that

(7.11) $\quad \rho_j \circ \gamma = q(j) \circ f \quad$ and $\quad \rho_k \circ \gamma = q(k) \circ g.$

Then there is $h : X \longrightarrow Q_0(j,k)$ so

(7.12) $\quad \rho_j \circ h = f \quad$ and $\quad \pi \circ h = \gamma.$

Now

(7.13) $\quad q(k) \circ \rho_k \circ h = \rho_k \circ \pi \circ h = q(k) \circ g$

$\Rightarrow \quad \rho_k \circ h = g$

because $q(k)$ is monomorphic. □

Proposition 7.14: Let $\mathfrak{C}$ be a topologized category. Let $b : A \longrightarrow B$ and $c : B \longrightarrow C$ be $\mathfrak{C}$-morphisms such that

(7.15.a) $\quad c$ and $c \circ b$ are flush local subsets,

(7.15.b) $\quad c$ is monomorphic.

Then b is a flush local subset.

Proof: Let S and T be covers of A and B, respectively, such that each of $\{c \circ t : t \in T\}$ and $\{c \circ b \circ s : s \in S\}$ covers C. For $(s,t) \in S \times T$, let $(s \times t; \pi_s, \pi_t)$ be a fibered product $(b \circ s) \times_B t$. As c is monomorphic, $(s \times t; \pi_s, \pi_t)$ is also a fibered product $(c \circ b \circ s) \times_C (c \circ t)$. Consequently, for each $t \in T$, the function $s \longmapsto (s \times t, \pi_t)$ on S is a cover of $\mathrm{dom}(t)$. It follows easily that $(s,t) \longmapsto (s \times t; s \circ \pi_s)$ on $S \times T$ is a cover of A whose composition with b is a cover of B. □

§8 Graph Reductions

Most of the following results are phrased in the context of a Grothendieck topology. As every category admits a default topology, the theorems also describe limits for an arbitrary category.

Definition 8.1: Let $\mathcal{D}$ be a category.

Let (S,M) be a graph and let G be a graph of $\mathcal{D}$-objects of type (S,M). Let $\alpha : G \longrightarrow A$ be a cone of G. We say that α is pseudosurjective if for each $B \in \mathcal{D}$, the function

(8.2) $\quad f \longmapsto (s \mapsto f \circ \alpha(s) \quad \text{on } s \in S)$

is a monomorphism from $\mathrm{Mor}_{\mathcal{D}}(A,B)$ to the set of cones from G to B. A source of G which satisfies the condition dual to this is called pseudoinjective.

For the remainder of this definition, assume $\mathcal{D}$ is a category with Grothendieck topology. Let $B_0 \in \mathcal{D}^p$ and $A \in \mathcal{D}$. There is a bijection Θ from cones $B_0 \longrightarrow A$ to $\mathrm{Mor}_{\mathcal{D}^p}(B_0, A^p)$ determined by

(8.3) $\quad$ for $j \in \Lambda(B_0)$ and $\beta : B_0 \longrightarrow A$ a cone, $\Theta[\beta](j) = \beta(j)^p$.

We freely identify these spaces, and freely apply adjectives applicable to either morphism and cones.

Suppose $b_0 : B_0 \longrightarrow A_0$ is a $\mathcal{D}^p$-morphism. We call b_0 a graph reduction if for each $C \in \mathcal{D}$, the function

(8.4) $\quad f \longmapsto f \circ b_0 \qquad$ from $\mathrm{Mor}_{\mathcal{D}^p}(A_0, C^p) \longrightarrow \mathrm{Mor}_{\mathcal{D}^p}(B_0, C^p)$

is bijective. A colimit may be regarded as a graph reduction to an affine object. If b_0 is a pullback base and every pullback of b_0 along a $\mathcal{D}^p$-morphism is a graph reduction, then we say b_0 is an absolute reduction.

Suppose $B, C \in \mathcal{C}$, $B_0, C_0 \in \mathcal{C}$, $b^\# : B_0 \longrightarrow B^p$ and $c^\# : C_0 \longrightarrow C^p$ are graph reductions, and $f_0 \in \mathrm{Mor}_{\mathcal{C}^p}(B_0, C_0)$. There is a unique $f \in \mathrm{Mor}_{\mathcal{C}}(B,C)$ such that $c^\# \circ f_0 = f^p \circ b^\#$, which is called the reduction of f_0 through $b^\#$ and $c^\#$.

Our objective is to characterize the canopy of an absolute cover as a special type of morphism. In fact, the projection of such a canopy is a monomorphic absolute reduction. However, as Definition 8.1 involves pullbacks along arbitrary $\mathcal{D}^p$-morphisms, this claim requires proof even when the topology is intrinsic. For now, we derive consequences of the definition.

For the remainder of this section, assume

(8.5) $(\mathfrak{C},\mathrm{Sub},\mathrm{Cov})$ is a topologized category.

Lemma 8.6: Assume (8.5).

(A) The class of absolute reductions in $\mathfrak{C}^p$ forms a universe of subsets.

(B) Suppose $c_0: C_0 \longrightarrow B_0$ and $b_0: B_0 \longrightarrow A_0$ are $\mathfrak{C}^p$-morphisms. If c_0 and $c_0 \circ b_0$ are absolute reductions and b_0 is a pullback base, then b_0 is an absolute reduction.

(C) Suppose $B_0 \in \mathfrak{C}^p$, $B \in \mathfrak{C}$ and $\Theta: B_0 \longrightarrow B$ is a cone. If Θ, regarded as a $\mathfrak{C}^p$-morphism, is a graph reduction, then Θ is a colimit in $\mathfrak{C}$.

Proof: Trivial.

Proposition 8.7: An absolute reduction is a $\mathfrak{C}^p$-epimorphism.

Proof: Suppose $b_0: B_0 \longrightarrow C_0$ is an absolute reduction. Obviously, if $\iota_j^{-1}(B_0,b_0)$ is epimorphic for each $j \in \Lambda(C_0)$, then b_0 is epimorphic. We are reduced to the case in which C_0 is affine.

Let $C \in \mathfrak{C}$ and $B_0 \in \mathfrak{C}^p$, and let $b_0: B_0 \longrightarrow C^p$ be an absolute reduction. Suppose $X_0 \in \mathfrak{C}^p$ and $f_0, g_0 \in \mathrm{Mor}_{\mathfrak{C}^p}(C^p, X_0)$ so that $f_0 \circ b_0 = g_0 \circ b_0$. Fix representatives $(f,j) \in f_0$ and $(g,k) \in g_0$. For $z \in \Lambda(B_0)$,

(8.8) $[f \circ b_0(z), j] = [g \circ b_0(z), k]$,

and there is a unique $h(z): B_0(z) \longrightarrow X_0(j,k)$ such that

(8.9) $\rho_j \circ h(z) = f \circ b_0(z)$ and $\rho_k \circ h(z) = g \circ b_0(z)$.

Using property (3.5.b) for X_0, it is simple to show that $h_0: z \mapsto [h(z),1]$ is a $\mathfrak{C}^p$-morphism $B_0 \longrightarrow X_0(j,k)^p$. Let h_1 be the factoring of h through b_0. Then

(8.10) $\forall z \in \Lambda(B_0)$, $\rho_j \circ h_1 \circ b_0(z) = f \circ b_0(z)$ and $\rho_k \circ h_1 \circ b_0(z) = g \circ b_0(z)$

$\Rightarrow$ $\rho_j \circ h_1 = f$ and $\rho_k \circ h_1 = g$

$\Rightarrow$ $f_0 = g_0$. □

Proposition 8.11: Let (S,M) be a graph. For each $j \in \{1,2\}$, let G_j be a graph of $\mathfrak{C}^p$-objects of type (S,M), and let $\Theta_j : B_j \longrightarrow G_j$ be a source. Let φ be a graph transformation $G_1 \longrightarrow G_2$ such that $\varphi(s)$ is monomorphic for each $s \in S$. Suppose $\beta : B_1 \longrightarrow B_2$ is an absolute reduction such that

(8.12) $\quad \forall s \in S, \quad \Theta_2(s) \circ \beta = \varphi(s) \circ \Theta_1(s).$

Suppose that Θ_1 is pseudoinjective.

(A) Θ_2 is pseudoinjective.

(B) Suppose

(8.13.a) $\quad \Theta_1$ is an inverse limit,

(8.13.b) $\quad B_2$ is affine,

(8.13.c) $\quad$ there is a finite subset $T \subseteq S$ such that for each $s \in S$ there is $t \in T$ so $M(t,s) \neq \varnothing$,

(8.13.d) $\quad \varphi(s)$ is an absolute reduction for each $s \in S$.

Then Θ_2 is an inverse limit. In particular, if G_3 is a graph of $\mathfrak{C}$-objects of type (S,M), and $\Theta_3 : B_3 \longrightarrow G_3$ is a source of $\mathfrak{C}$-objects so that $G_2 = G_3{}^p$ and $\Theta_2 = \Theta_3{}^p$, then Θ_3 is an inverse limit in $\mathfrak{C}$.

Proof: (A) Let $X \in \mathfrak{C}^p$ and let $f,g : X \longrightarrow B_2$ such that $\Theta_2(s) \circ f = \Theta_2(s) \circ g$ for each $s \in S$. Let $(X_f;\pi_X,\pi_1)$ and $(X_g;\pi_X,\pi_1)$ be pullbacks of β along f and g, respectively. Let $(X_0;\pi_f,\pi_g)$ be a fibered product $X_f \times_X X_g$, and then

(8.14) $\quad \gamma = \pi_X \circ \pi_f = \pi_X \circ \pi_g$

is an absolute reduction.

Let $s \in S$. Then

(8.15) $\quad \varphi(s) \circ \Theta_1(s) \circ \pi_1 \circ \pi_f = \Theta_2(s) \circ \beta \circ \pi_1 \circ \pi_f = \Theta_2(s) \circ f \circ \gamma.$

Similarly, $\varphi(s) \circ \Theta_1(s) \circ \pi_1 \circ \pi_g = \Theta_2(s) \circ g \circ \gamma$. But $\varphi(s)$ is monomorphic. Hence

(8.16) $\quad \forall s \in S, \quad \Theta_1(s) \circ \pi_1 \circ \pi_f = \Theta_1(s) \circ \pi_1 \circ \pi_g$

$\Rightarrow \quad \pi_1 \circ \pi_f = \pi_1 \circ \pi_g$

$\Rightarrow \quad f \circ \gamma = \beta \circ \pi_1 \circ \pi_f = \beta \circ \pi_1 \circ \pi_g = g \circ \gamma.$

As γ is epimorphic, $f = g$.

(B) Assume (8.13.a,b,c), and suppose $D \in \mathfrak{C}$ and $\delta : D^p \longrightarrow G_2$ is a source. For each $t \in T$, let $(D_t;\pi_D,\pi_t)$ be a pullback $\delta(t)^{-1}(G_1(t),\varphi(t))$. As T is finite, a simple recursion establishes existence of a product $(D_0;\varepsilon)$ of the family $\{(D_t,\pi_D)\}_{t \in T}$ in $\mathfrak{C}^p/D^p$ such that $\varepsilon(t)$ is an absolute reduction for each $t \in T$. Define γ to be $\pi_D \circ \varepsilon(t)$ for every $t \in T$.

For $s \in S$, $t \in T$ and $(t,\rho,s) \in M$,

$$(8.17)\qquad \varphi(s)\circ G_1(t,\rho,s)\circ\pi_t\circ\varepsilon(t) = G_2(t,\rho,s)\circ\varphi(t)\circ\pi_t\circ\varepsilon(t)$$
$$= G_2(t,\rho,s)\circ\delta(t)\circ\pi_D\circ\varepsilon(t)$$
$$= \delta(s)\circ\gamma.$$

Since φ consists of monomorphisms, we conclude existence of a source $\delta' : D_0 \longrightarrow G_1$ uniquely determined by the property that $\varphi(s)\circ\delta'(s) = \delta(s)\circ\gamma$ for $s \in S$. Let d' be the unique factoring of δ' through B_1. As B_2 is affine, there is a unique $d_2 : D^p \longrightarrow B_2$ such that $d_2\circ\gamma = \beta\circ d'$. Clearly, d_2 factors δ through Θ_2. □

<u>Corollary 8.18:</u> Let $b_0 : B_0 \longrightarrow C_0$ be an absolute reduction in $\mathfrak{C}^p$.

Let σ be a function which to each $x \in \mathrm{dom}(\sigma)$ assigns a $\mathfrak{C}^p$-morphism $s(x) : C_0 \longrightarrow S(x)$. Assume that if $X_0 \in \mathfrak{C}^p$ and $f_0, g_0 \in \mathrm{Mor}_{\mathfrak{C}^p}(X_0, B_0)$ such that $s(x)\circ b_0\circ f_0 = s(x)\circ b_0\circ g_0$ for each $x \in \mathrm{dom}(\sigma)$, then $f_0 = g_0$. In this case, if $X_0 \in \mathfrak{C}^p$ and $f_0, g_0 \in \mathrm{Mor}_{\mathfrak{C}^p}(X_0, C_0)$ so $s(x)\circ f_0 = s(x)\circ g_0$ for each $x \in \mathrm{dom}(\sigma)$, then $f_0 = g_0$.

In particular, if $c_0 : C_0 \longrightarrow D_0$ is a $\mathfrak{C}^p$-morphism such that $c_0\circ b_0$ is monomorphic, then c_0 is monomorphic.

<u>Proof:</u> The properties with respect to σ are expressable as pseudoinjectivity of certain sources. □

§9 Absolute Covers and Limits in $\mathcal{C}^P$

We continue under hypothesis (8.5). It is time to develop the many properties of assigned covers.

Proposition 9.1: Let $\mathcal{D}$ be a category. Let (S,M) and (T,N) be graphs, let A_1 and B_1 be graphs of $\mathcal{D}$-objects of types (S,M) and (T,N), respectively, and let $\alpha : A_1 \longrightarrow A$ and $\beta : B_1 \longrightarrow B$ be cones. Let $b \in \mathrm{Mor}_{\mathcal{D}}(B,A)$. Suppose that

(9.2.a) a pullback of α along b exists,

(9.2.b) for each $t \in T$, a pullback of α along $(b \circ \beta(t))$ exists and is a colimit,

(9.2.c) β is a colimit, and

(9.2.d) for each $s \in S$, the pullback of β along $\{b^{-1}\alpha\}(s)$ is a colimit.

Then the pullback of α along b is a colimit.

Proof: By an obvious diagram chase. □

Corollary 9.3: Let $\mathcal{C}$ be a topologized category. For $B_0 \in \mathcal{C}^P$, the assigned cover of B_0 is an absolute cover.

Proof: First, suppose that $c_0 : C_0 \longrightarrow A_0$ is a $\mathcal{C}^P$-morphism such that $c_0^{-1}\iota_j$ is affine for each $j \in \Lambda(A_0)$. Theorem 4.13 applies, and consequently the pullback of the assigned cover is a colimit.

Suppose $b_0 : B_0 \longrightarrow A_0$ is any $\mathcal{C}^P$-morphism. Let $\alpha : A_1 \longrightarrow A_0$ and $\beta : B_1 \longrightarrow B_0$ be the cones of the corresponding assigned covers. For $j \in \Lambda(B_0)$ and $k \in \Lambda(A_0)$, $(b_0 \circ \beta(j)) \times_A \alpha(k)$ is affine by Proposition 3.33. The previous paragraph implies that these graphs meet the hypothesis of Proposition 9.1. □

Corollary 9.4: Let $\mathcal{C}$ be a topologized category, and let $b_0 : B_0 \longrightarrow A_0$ be a pullback base for $\mathcal{C}^P$. Suppose the pullback of b_0 along every affine morphism into A_0 is graph reduction. Then b_0 is an absolute reduction.

Proof: Let $c_0 : C_0 \longrightarrow A_0$ be a $\mathcal{C}^p$-morphism, and regard the assigned cover of C_0 as an absolute cover. A diagram chase yields that the pullback is a graph reduction. □

Definition 9.5: Let $\mathcal{C}$ be a topologized category. Let $B_0 \in \mathcal{C}^p$ and let Θ_0 be a function which to each $j \in \operatorname{dom}\Theta_0$ assigns $D(j) \in \mathcal{C}$ and $d(j) \in \operatorname{Sub}^p(D(j)^p, B_0)$. Let Θ_1 denote the function $j \mapsto (D(j)^p, d(j))$ on $\operatorname{dom}(\Theta_0)$. We call Θ_0 an explicitly affine cone into B_0, and frequently identify it with Θ_1. For $(j,k) \in \operatorname{dom}(\Theta_0)^2$, there is a triple

(9.6) $D(j,k) \in \mathcal{C}$,

$\rho_1[j,k] \in \operatorname{Mor}_{\mathcal{C}}(D(j,k), \operatorname{dom}\Theta(j))$,

$\rho_2[j,k] \in \operatorname{Mor}_{\mathcal{C}}(D(j,k), \operatorname{dom}\Theta(k))$

such that $(D(j,k)^p; \rho[j,k](1)^p, \rho[j,k](2)^p)$ is a fibered product $\Theta_1(j) \times_{B_0} \Theta_1(k)$. Let D_0 be the intersection graph of $\mathcal{C}$-objects on $\operatorname{Int}(\operatorname{dom}(\Theta_0))$ given by

(9.7) $D_0(j) = D(j)$ for $j \in \operatorname{dom}(\Theta_0)$,

$D_0(j,k) = D(j,k)$ for $j,k \in \operatorname{dom}(\Theta_0)$,

$D_0((j,k), \rho_1, j) = \rho_1[j,k]$ and

$D_0((j,k), \rho_2, k) = \rho_2[j,k]$ for $j,k \in \operatorname{dom}(\Theta_0)$.

We refer to D_0 as a pushdown canopy of Θ_0; moreover, $j \mapsto \Theta_1(j)$ is a $\mathcal{C}^p$-morphism to which we refer as canonical. The image of the graph of D_0 under the pasting functor is a choice of canopy for Θ_1 with respect to the category $\mathcal{C}^p$.

Let Θ_0 and B_0 be as before. Suppose $c_0 : C_0 \longrightarrow B_0$ is a $\mathcal{C}^p$-morphism, and suppose φ is a function which to each $j \in \operatorname{dom}(\Theta_0)$ assigns

(9.8) $P(j) \in \mathcal{C}$, $\pi_D(j) \in \operatorname{Mor}_{\mathcal{C}}(P(j), D(j))$, $\pi_C(j) \in \operatorname{Mor}_{\mathcal{C}^p}(P(j)^p, C_0)$

such that

(9.9) $(P(j)^p; \pi_C(j), \pi_D(j)^p)$ is a fibered product $(C_0, c_0) \times_{B_0} \Theta_1(j)$.

We refer to either φ or to the function $j \mapsto (P(j), \pi_C(j))$ as an explicitly affine pullback. Obviously such choices are unique up canonical $\mathcal{C}$-isomorphisms. In particular, if C_0 is affine, such pullbacks exist.

Theorem 9.10: Let $\mathcal{C}$ be a topologized category, let $B_0 \in \mathcal{C}^p$, and let Θ be an explicitly affine cone of formal $\mathcal{C}^p$-subsets of B_0. Let $c_0 : C_0 \longrightarrow B_0$ be a pushdown canopy of Θ. Suppose

(9.11.a) c_0 is a pullback base in $\mathcal{C}^p$, and

(9.11.b) for $j \in \Lambda(B_0)$, any explicitly affine pullback of Θ along ι_j is an absolute cover.

Then c_0 is a monomorphic absolute reduction. In particular, if C_0, regarded as a graph of $\mathcal{C}$-objects, admits a colimit $\gamma : C_0 \longrightarrow C$, then there is a unique cone $\gamma' : B_0 \longrightarrow C$ such that $\gamma = \gamma' \circ c_0$, and this γ' is a colimit.

Proof: Proposition 7.7 implies that c_0 is monomorphic.

Let $D \in \mathcal{C}$ and let $d_0 : D^p \longrightarrow B_0$ be a $\mathcal{C}^p$-morphism. Express d_0 as $[f,j]$ for $j \in \Lambda(B_0)$ and $f \in \mathrm{Mor}_{\mathcal{C}}(D, B_0(j))$. Theorem 4.13 characterizes the pullback of c_0 along d_0 as the canopy of any explicitly affine pullback of Θ along d_0. However, the pullback along ι_j is given as an absolute cover; consequently, $d_0^{-1}c_0$ is a graph reduction. Now Corollary 9.4 implies that c_0 is an absolute reduction. □

Remark 9.12: Actually, condition (9.11.a) in Lemma 9.10 is redundant. Alas, we need the weaker form to develop the stronger.

There is a case of immediate interest. Let $B_0 \in \mathcal{C}^p$, and for each $j \in \Lambda(B_0)$ let φ_j be an absolute cover of $B_0(j)$. Let Θ be the subdivision of the assigned cover by $j \longmapsto p \circ \varphi_j$, where p is the pasting functor. Let $c_0 : C_0 \longrightarrow B_0$ be a pushdown canopy. We claim that (9.11.a,b) hold.

Clearly c_0 is a formal pullback base. For $j,k \in \Lambda(B_0)$ and $x \in \mathrm{dom}(\varphi_k)$, pullback of $\varphi_k(x) \circ \iota_k$ along ι_j is composition of of pullback of $\varphi_k(x)$ along ρ_k on $B_0(j,k)$ with ρ_j. To show (9.11.b), it suffices to show that, for $j \in \Lambda(B_0)$, ρ_1 and ρ_2 on $B_0(j,j)$ are covering morphisms with respect to the *default* topology. Proposition 2.10 proves this. □

Theorem 9.13: Let B_1 be a canopy of $\mathcal{C}^p$-objects (with respect to the $\mathcal{C}^p$-topology). Then there is $C_0 \in \mathcal{C}^p$ and an indexed family $\Theta : \Lambda(B_1) \longrightarrow \mathcal{C}^p / C_0$ such that

(9.14.a) Θ is a cover of C_0 in the $\mathcal{C}^p$-topology,

(9.14.b) Θ is an absolute cover of C_0 with respect to $\mathcal{C}^p$,

(9.14.c) B_1 is a canopy of Θ.

In particular, there is a monomorphic absolute reduction $B_1 \longrightarrow C_0$.

Proof: There are two categories with topologies here. We use suffixes $\mathcal{C}$ and $\mathcal{D} \equiv \mathcal{C}^p$ to emphasize category.

Let

(9.15) $\Lambda = \{(j,r) : j \in \Lambda(B_1),\ r \in \Lambda(B_1(j))\}$.

For (j,r), let $C(j,r) = \{B_1(j)\}(r)$, and let $\tau_{j,r} = \iota_j \circ (\iota_r{}^P)$ on $C(j,r)^{PP}$. Let τ denote $(j,r) \mapsto \tau_{j,r}$ on Λ. The previous observation states that τ meets the hypothesis of Theorem 9.10. Let $c_1 : C_1 \longrightarrow B_0$ be the pushdown formalization, and then c_1 is a monomorphic absolute reduction with respect to $\mathcal{D}$.

For $((j,r),(k,s)) \in \Lambda$, the projections on $C_1((j,r),(k,s))$ are formal $\mathcal{D}$-subsets into $\mathfrak{C}$-affine objects. This means that $C_1((j,r),(k,s))$ can be identified with a $\mathfrak{C}$-object in such a manner that the projections become formal $\mathfrak{C}$-subsets. Without loss of generality, assume C_0 is a canopy of $\mathfrak{C}$-objects so that for $\Lambda(C_0) = \Lambda(C_1)$ and $C_1(\beta) = C_0(\beta)^P$ for each β in the object or morphism sets of $\mathrm{Int}(\Lambda(C_0))$. There is a morphism $c^{\#} : C_1 \longrightarrow C_0{}^P$ given by $j \mapsto \iota_j{}^P$.

There is $\gamma : B_1 \longrightarrow C_0{}^P$ so that $\gamma \circ c_1 = c^{\#}$. Put $\Theta(j) = \gamma \circ \iota_j$ for $j \in \Lambda(B_1)$. By Corollary 8.18, γ is monomorphic. For $j \in \Lambda(B_1)$ and $(k,s) \in \Lambda$, pullback of $\iota_{k,s} = \gamma \circ c_1 \circ \iota_{k,s} = \gamma \circ \tau_{k,s}$ along $\Theta(j)$ is the pullback of $\tau_{k,s}$ along the j-th coordinate chart of B_1. In rapid succession, we may conclude

(9.16.a) $\Theta(j)$ is a formal $\mathfrak{C}^P$-subset for each index j,

(9.16.b) B_1 is a canopy of Θ,

(9.16.c) the assigned cover of C_0 is a subcover of Θ.

Observations (9.16.a,c) imply that Θ is an absolute cover in $\mathfrak{C}^P$. Thus, γ is a pullback base in $\mathfrak{C}^P$. Lemma 8.6 applies. □

<u>Corollary 9.17:</u> Let $A_0 \in \mathfrak{C}^P$ and let $(B_0,b_0),(C_0,c_0) \in \mathfrak{C}^P/A_0$. If for each $j \in \Lambda(C_0)$, the pullback of b_0 along $c_0 \circ \iota_j$ exists, then the pullback of b_0 along c_0 exists.

<u>Proof:</u> Put $\mathcal{D} = \mathfrak{C}^P$. Let $c_1 : C_1 \longrightarrow C_0{}^P$ denote a canopy of the assigned cover to C_0. Apply Theorem 4.13 to canopies in $\mathcal{D}$ to deduce existence of a fibered product $(P_1;\pi_{B,1},\pi_{C,1})$ of $B_0{}^P \times_{A_0{}^P} (C_1, c_0{}^P \circ c_1)$. By the previous theorem, there is a monomorphic absolute reduction $\tau^{\#} : P_1 \longrightarrow P_0{}^P$ where $P_0 \in \mathcal{D}$. Let $\pi_{B,0}$ and $\pi_{C,0}$ be the unique $\mathcal{D}$-morphisms on P_0 so

(9.18) $\pi_{B,0}{}^{P} \circ \tau^{\#} = \pi_{B,1}$ and $\pi_{C,0}{}^{P} \circ \tau^{\#} = c_1 \circ \pi_{C,1}$.

The situation satisfies the hypothesis of Proposition 8.11.B with respect to the ground category $\mathcal{D}$. Thus, $(P_0;\pi_{B,0},\pi_{C,0})$ is a fibered product in $\mathcal{D}$. □

Corollary 9.19: Let $b_0 : B_0 \longrightarrow A_0$ be a $\mathfrak{C}^P$-morphism. If for each $j \in \Lambda(A_0)$ the pullback of b_0 along ι_j is a pullback base, then b_0 is a pullback base.

Definition 9.20: Assume (8.5). For $B_0 \in \mathfrak{C}^P$, an affinization of B_0 is a cone $\beta : B_0 \longrightarrow B$ such that, for $\beta_0 = \beta|_{\Lambda(B_0)}$,

(9.21.a) β_0 is a cover of B in the topology of $\mathfrak{C}$,

(9.21.b) β_0 is an absolute cover of B,

(9.21.c) B_0 is a canopy derived from β_0.

Then B_0 is called pseudoaffine. A reduction of a morphism through affinizations of domain and codomain is also referred to as an affinization.

We say $\mathfrak{C}$ is closed under affinization if every canopy of $\mathfrak{C}$ has an affinization. In particular, if $\mathfrak{C}$ is a topologized category, then $\mathfrak{C}^P$ is closed under affinization.

Remark 9.22: For an arbitrary topologized category $\mathfrak{C}$, composition of an affinization with a $\mathfrak{C}^P$-isomorphism need not be an affinization. Moreover, the notion of affinization must be extended not only to morphisms into members of $\mathfrak{C}$, but also to morphisms between different canopies. 'Pseudoisomorphisms', introduced in Section 12, will extend the concept.

Corollary 9.17 yields stronger versions of Proposition 9.1 and Lemma 9.10.

Proposition 9.23: Let $\mathfrak{C}$ be a topologized category. Let G be a graph of $\mathfrak{C}^P$-objects and let $\beta : G \longrightarrow B_0$ be a cone.

(A) Suppose a pullback of β exists along any affine morphism into B_0. Then a pullback of β along any $\mathfrak{C}^P$-morphism exists.

(B) Suppose for b an affine morphism into B_0, a pullback of β along b exists and such a pullback is a colimit. Then for b any $\mathfrak{C}^P$-morphism into B_0, a pullback of β along b exists and such a pullback is a colimit.

Proposition 9.24: Let $\mathfrak{C}$ be a topologized category, let $B_0 \in \mathfrak{C}^p$, and let Θ be an explicitly affine cone of formal $\mathfrak{C}^p$-subsets of B_0. Let $c_0 : C_0 \longrightarrow B_0$ be a pushdown canopy of Θ. Suppose that for $j \in \Lambda(B_0)$, any explicitly affine pullback of Θ along ι_j is an absolute cover. Then c_0 is a monomorphic absolute reduction.

§10 Pullback Systems

The following example motivates the author's introduction of pullback systems in Definition 4.7. Let **Ring⁰** be the opposite category of rings, let **Sch** be the category of schemes, and let $\Phi : \mathbf{Ring}^0 \longrightarrow \mathbf{Sch}$ be the canonical functor.

(10.1.a) A covering morphism of $\mathbf{Ring}^0$ is an isomorphism. This property is true in **Sch** the 'globalization' of $\mathbf{Ring}^0$. In particular, if $A \in \mathbf{Ring}^0$, then a covering morphism of $\Phi(A)$ is $\mathbf{Sch}/\Phi(A)$-isomorphic to Φ of a member of $\mathbf{Ring}^0/A$.

(10.1.b) Consider Ét the class of étale surjections in $\mathbf{Ring}^0$. It is natural to call a Sch-morphism b an étale surjection if there is an affine cover of cod b each of whose members pulls back to an étale surjection in the sense of $\mathbf{Ring}^0$. What is less obvious, but still true, is that any étale surjection in **Sch** onto $\Phi(A)$, for any $A \in \mathbf{Ring}^0$, is isomorphic to an étale surjection in $\mathbf{Ring}^0/A$! Classically, the category **Sch** is created by embedding $\mathbf{Ring}^0$ into the category of sheaves in a clever way; with this outlook, the fact that étale surjections to affines are just those from $\mathbf{Ring}^0$ appears as a convenient accident. Our present effort claims that Sch is, in fact, intrinsically determined by $\mathbf{Ring}^0$. But then, the above condition must be the consequence of the mapping properties of Ét.

Let us return to the language of canopies. Consider a $\mathbf{Ring}^{0\mathrm{p}}$-morphism $b_0 : B_0 \longrightarrow A_0$; again, imagine that b_0 represents a Sch-morphism. For b_0 to represent an étale surjection of **Sch**, there must be an explicitly affine $\mathbf{Ring}^{0\mathrm{p}}$-cover of A_0 each of whose pullbacks represents an Ét-morphism.

Consider the general case of a topologized category $\mathfrak{C}$ and a universe Lay of layered morphisms for $\mathfrak{C}$. The universe of special interest is $\mathrm{Lay} = \mathrm{Cvm}_{\mathfrak{C}}$. The basis of our method for extending Lay is elementary: A $\mathfrak{C}^{\mathrm{p}}$-morphism $b_0 : B_0 \longrightarrow A_0$ is a 'Lay_1'-morphism if there is any $\mathfrak{C}^{\mathrm{p}}$-cover of A_0 each of whose pullbacks along b_0 represents a Lay-morphism. *The process will not work for an arbitrary universe of layered morphisms.* To specify which universes are susceptible the approach, formulate

Definition 10.2: Let $\mathfrak{C}$ be a category with Grothendieck topology. Let Lay be a universe of layered morphisms for $\mathfrak{C}$.

Let $C_0 \in \mathfrak{C}^{\mathrm{p}}$ and let (Q_0,q) be a pullback system of C_0. We say (Q_0,q) is a pullback system of Lay-morphisms if $q(j)$ is a Lay-morphism for each $j \in \Lambda(F)$. We say that $\mathfrak{C}$ is closed under Lay if for C_0 a pseudoaffine object in $\mathfrak{C}^p$ and for (Q_0,q) a pullback system of Lay-morphisms into C_0, the object Q_0 is pseudoaffine and the affinization of $j \mapsto [q(j),j]$ is a Lay-morphism.

Even the above condition is inadequate without some assumptions on the topology. Consequences of closure are postponed until Chapter 15. In this section, we establish some basic facts about pullback systems.

Proposition 10.3: Let $A \in \mathfrak{C}$, $(B,b),(C,c) \in \mathfrak{C}/A$, and let $c^{\#} : C_0 \longrightarrow C^p$ be an affinization. Suppose (Q_0,q) is a pullback system for $c^p \circ c^{\#}$ along b^p, and let $\pi_0 : Q_0 \longrightarrow B$ denote the canonical morphism. Suppose that $q^{\#} : Q_0 \longrightarrow Q^p$ is an absolute reduction. Let q_0 denote $j \mapsto [q(j),j]$ on $\Lambda(Q_0)$. Let π_B and π_C denote the affinizations of π_0 and of q_0, respectively, through $q^{\#}$ and $c^{\#}$. Then $(Q_0;\pi_B,\pi_C)$ is a pullback $b^{-1}(C,c)$.

Proof: Theorem 4.13 states that Q_0 is a fibered product of $B^p \times_{A^p} C_0$. Invoke Proposition 8.11. □

Corollary 10.4: Suppose $\mathfrak{C}$ is a topologized category which is closed under affinization. Let $b \in \mathrm{Mor}(\mathfrak{C})$ and let S be a cover of dom b such that

(10.5.a) for each $s \in S$, $b \circ s$ is a pullback base,

(10.5.b) S is an absolute cover of dom b.

Then b is a pullback base.

Proof: Trivial. □

Corollary 10.6: Let Lay be a system of layered morphisms for $\mathfrak{C}$, and suppose $\mathfrak{C}$ is closed under Lay. Let $A \in \mathfrak{C}$, $(B,b),(C,c) \in \mathfrak{C}$ and let S be a cover of C. Suppose

(10.7.a) a canopy map of S is an affinization,

(10.7.b) for each $s \in S$, there is a pullback $(c \circ s)^{-1}(B,b)$ which is a Lay-morphism.

Then there is a pullback $c^{-1}(B,b)$ which is a Lay-morphism.

Abstract pullback systems always arise as pullbacks along something.

<u>Proposition 10.8:</u> Let $C_0 \in \mathcal{C}^P$ and let (Q_0,q) be a pullback system into C_0. Let q_0 denote $j \mapsto [q(j),j]$. For $j \in \Lambda(C_0)$, let $\iota_{j,C}$ and $\iota_{j,Q}$ denote the j-th coordinate charts of C_0 and Q_0, respectively. Then for $j \in \Lambda(C_0)$, $(Q_0(j)^P;\iota_{j,Q},q(j)^P)$ is a pullback $q_0^{-1}\iota_{j,C}$.

<u>Proof:</u> For $j.k \in \Lambda(C_0)$, let $q(j,k) : Q_0(j,k) \longrightarrow C_0(j,k)$ be the unique morphism such that $\rho_j \circ q(j,k) = \rho_j$ and $\rho_k \circ q(j,k) = \rho_k$.

Fix $j \in \Lambda(C_0)$ and $D \in \mathcal{C}$. Suppose $f_0 \in \mathrm{Mor}_{\mathcal{C}^P}(D^P,Q_0)$ and $g \in \mathrm{Mor}_{\mathcal{C}}(D,C_0(j))$ such that $\iota_j \circ g^P = q_0 \circ f_0$. Fix a representative $(f,k) \in f_0$. There is $h_1 \in \mathrm{Mor}_{\mathcal{C}}(D,C_0(j,k))$ so

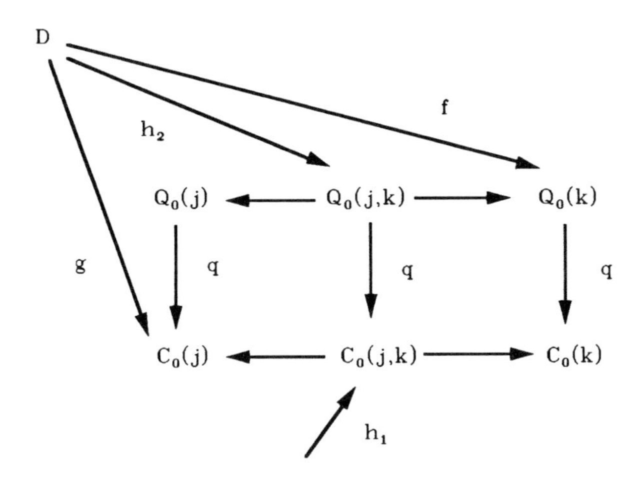

(10.9) $\qquad \rho_j \circ h_1 = g \quad \text{and} \quad \rho_k \circ h_1 = q(k) \circ f.$

By hypothesis, there is $h_2 : D$ Æ$\to Q_0(j,y)$ such that

(10.10) $\qquad \rho_k \circ h_2 = f \quad \text{and} \quad q(j,k) \circ h_2 = h_1.$

Put $h = \rho_j \circ h_2$.

Now,

(10.11) $\qquad q(j) \circ h = \rho_j \circ q(j,k) \circ h_2 = \rho_j \circ h_1 = g,$

$\qquad \qquad \iota_j \circ h^P = [h,j] = [f,k] = f_0.$

Next, suppose $d : D \longrightarrow Q_0(j)$ such that $q(j) \circ d = g$ and $\iota_j \circ d^P = f_0$. As $[f,k] = [d,j]$, there is $d_2 : D \longrightarrow Q_0(j,y)$ so that $\rho_k \circ d_2 = f$ and $\rho_j \circ d_2 = d$. Thus,

(10.12) $\rho_j \circ q(j,k) \circ d_2 = q(j) \circ \rho_j \circ d_2 = g = \rho_j \circ h_1,$

$\rho_k \circ q(j,k) \circ d_2 = q(k) \circ \rho_k \circ d_2 = q(k) \circ f = \rho_k \circ h_1,$

$\Rightarrow \quad q \circ d_2 = h_1.$

But then $q(j,k) \circ d_2 = q(j,k) \circ d_2$ and $\rho_k \circ d_2 = \rho_k \circ h_2$, so $d_2 = h_2$. Thus, $d = h$. □

Corollary 10.13: Let $C_0 \in \mathfrak{C}^p$ and let (Q_0,q) be a pullback system for C_0. Let q_0 denote $j \mapsto [q(j),j]$ on $\Lambda(C_0)$. Suppose $C,Q \in \mathfrak{C}$ and $c^{\#}: C_0 \longrightarrow C^p$ and $q^{\#}: Q_0 \longrightarrow Q^p$ are affinizations. Let q_1 be the affinization of q_0 through $c^{\#}$ and $q^{\#}$. Then $(Q_0;q^{\#},q_0)$ is a pullback $(q_1{}^p)^{-1}(C_0,c^{\#})$.

Proof: By Theorem 4.13, it suffices to prove that for each $j \in \Lambda(C_0)$, $(Q_0(j)^p;q^{\#} \circ \iota_j,q(j)^p)$ is a pullback $(q_1{}^p)^{-1}(C_0(j)^p,b^{\#} \circ \iota_j)$. Propositions 8.11 and 10.8 assure this. □

We can now add a variant to Proposition 6.1. Begin with

Lemma 10.14: Let $\mathfrak{C}$ be a topologized category and let $f_0: B_0 \longrightarrow A_0$ be a $\mathfrak{C}^p$-morphism. Suppose B_0 and C_0 are pseudoaffine, and that $f \in \mathrm{Mor}(\mathfrak{C})$ is an affinization.

(A) Suppose f_0 is a pullback base in $\mathfrak{C}^p$ and every pullback along f_0 is pseudoaffine. Then f is a pullback base.

(B) Suppose the topology of $\mathfrak{C}$ satisfies the weak CLCS condition. If f_0 is a formal $\mathfrak{C}^p$-subset, then f is a formal $\mathcal{D}$-subset.

Proof: (B) By Proposition 10.8 and Lemma 2.46. □

Proposition 10.15: Let $(\mathfrak{C},\mathrm{Sub},\mathrm{Cov})$ be a topologized category and let $\Gamma: \mathfrak{C} \longrightarrow \mathcal{D}$ be a functor of sections. Let p_0 denote the pasting functor for $\mathfrak{C}$. Let (Sub',Cov') be a Grothendieck topology for $\mathcal{D}$. Assume (Sub',Cov') is an intrinsic topology which satisfies the CLCS condition. Let $\Phi: \mathfrak{C}^p \longrightarrow \mathcal{D}$ be a functor of sections such that $\Gamma = \Phi \circ p_0$. Suppose

(10.16.a) Γ is continuous,

(10.16.b) if G is a canopy over $\mathcal{C}$ and $\alpha : \Gamma(G) \longrightarrow A$ is a colimit in $\mathcal{D}$, then $\alpha' : j \mapsto \alpha(j)$ on $\Lambda(G)$ is an indexed cover with respect to (Sub',Cov') and $\Gamma(G)$ is a canopy with respect to α'.

(A) Let $A_0 \in \mathcal{C}^p$ and $(B_0,b_0),(C_0,c_0) \in \mathcal{C}^p/A_0$. Suppose that for each $j \in \Lambda(C_0)$, a fibered product for $(B_0,b_0)\times_{A_0}\{c_0 \circ \iota_j\}$ exists and is preserved by Φ (where ι_j is the j-th coordinate chart of C_0). Then a fibered product of $(B_0,b_0)\times_{A_0}(C_0,c_0)$ exists and is preserved by Φ.

(B) Φ maps formal $\mathcal{C}^p$-subsets to formal $\mathcal{D}$-subsets.

(C) If (Sub',Cov') is flush, then Φ is continuous.

<u>Proof:</u> Every canopy of a $\mathcal{D}$-cover and every canopy of the type in (10.16.b) can be interpreted as an affinization in $\mathcal{D}^p$. Consider $A_0 \in \mathcal{C}^p$ and $(B_0,b_0),(C_0,c_0) \in \mathcal{C}^p/A_0$ such that a pullback system (Q_0,q) of c_0 along b_0 exists (that is, the hypothesis of (A). Let $q^{\#} : Q_0 \longrightarrow Q$ be a colimit in $\mathcal{C}^p$. Previous lemmas assure that the fibered product $B_0\times_{A_0}C_0$ exists and is preserved by Φ provided that Φ maps Q_0 to the canopy of an absolute cover in $\mathcal{D}$. Thus, the proof of (A) is entwined with proof of (B) and (C).

This first remark assures that Φ preserves all pullbacks along any formal $\mathcal{C}^p$-subset from an affine object of $\mathcal{C}^p$. For b_0 a formal $\mathcal{C}^p$-subset, its image under Γ satisfies the hypothesis of Lemma 10.14; hence, $\Phi(b_0)$ is a formal $\mathcal{D}$-subset.

Let Θ be an indexed cover of some $B_0 \in \mathcal{C}^p$. Let $j \in \Lambda(B_0)$. Then $\iota_j^{-1}\Theta$ may be identified with a $\mathcal{C}$-cover of $B_0(j)$. Our first remark assures that Φ of $\iota_j^{-1}\Theta$ is a pullback of $\Phi\circ\Theta$ along $\Phi(\iota_j)$. By hypothesis, $j \mapsto \Phi(\iota_j)$ is a $\mathcal{D}$-cover (and so an absolute cover) of $\Phi(B_0)$. Hence, $\Phi\circ\Theta$ is a cone of pullback bases and an unrefined cover with respect to Cov'. The remaining claims of (A), (B) and (C) follow easily. □

PART IV SMOOTHING

§11 The Smoothing Functor

For this section, assume

(11.1.a) ($\mathcal{C}$,Sub,Cov) is a topologized category,

(11.1.b) $A \mapsto \text{Small}(A)$ is a categorical choice of representatives with topology for $\mathcal{C}$.

Formally, our constructions depend on the choice of Small. However, it will be clear that alternative choices lead to naturally equivalent results.

Suppose $A,B \in \mathcal{C}$. By a global map from A to B, we mean a pair (Θ,f) where

(11.2.a) Θ and f are functions on a common domain,

(11.2.b) Θ is an indexed cover of A,

(11.2.c) for each $x \in \text{dom}(\Theta)$, $f(x) \in \text{Mor}_{\mathcal{C}}(\text{dom}\,\Theta(x),B)$,

(11.2.d) for $x,y \in \text{dom}(\Theta)$ and $(P;\pi_x,\pi_y)$ a product for $\Theta(x)\times_A\Theta(y)$, $f(x)\circ\pi_x = f(y)\circ\pi_y$.

Suppose (Θ,f) is a global map from A to B and $d: D \longrightarrow A$ is a morphism which factors through Θ. Property (11.2.d) implies that there is a unique function $d_0 : D \longrightarrow B$ determined by the condition that

(11.3) for $x \in \text{dom}(\Theta)$ and $g \in \text{Mor}_{\mathcal{C}/A}((D,d),\Theta(x))$, $d_0 = f(x)\circ g$.

We refer to d_0 as $(\Theta,f)\circ d$. If η is an indexed family which factors through Θ, then the function $f_0(x) = (\Theta,f)\circ\eta(x)$ on $x \in \text{dom}(\Theta)$ is called a simple composition of (Θ,f) with η, and is denoted by $(\Theta,f)\circ\eta$. If η is a refinement of Θ—that is, a cover of A—then refer to either $(\Theta,f)\circ\eta$ or $(\eta,(\Theta,f)\circ\eta)$ as the refinement of (Θ,f) through η.

Let $A,B,C \in \mathcal{C}$ and let (φ,g) be a global map from B to C. If Θ_1 and Θ_2 are indexed families of $\mathcal{C}/B$ such that Θ_2 factors through Θ_1 and Θ_1 factors through φ, then tautologically

(11.4) $(\Theta_1,(\varphi,g)\circ\Theta_1)\circ\Theta_2 = (\varphi,g)\circ\Theta_2.$

Now suppose (Θ,f) is a global map from A to B such that f factors through φ. If η is an indexed family which factors through Θ, then

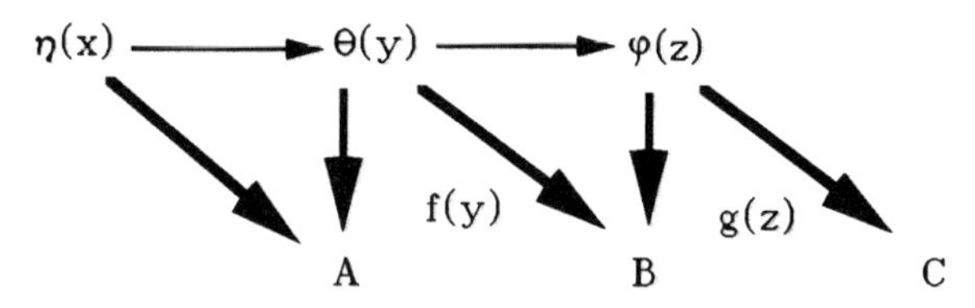

(11.5) $(\Theta,(\varphi,g)\circ f)\circ\eta = (\varphi,g)\circ\{(\Theta,f)\circ\eta\}$

as indicated by the diagram.

Proposition 11.6: Let $A,B \in \mathcal{C}$. Two global maps from A to B are called similar if they share a common refinement.

(A) A global map from A to B is similar to itself.

(B) Suppose α, β and γ are global maps. If α and β are similar, and β and γ are similar, then α and γ are similar.

(C) Define GM(A,B) to be the set of global maps (θ,f) for which θ is the identity function on a subset of Small(A). Then

(11.7.a) Every global map from A to B is similar to a member of GM(A,B).

(11.7.b) Two similar global maps from A to B share a common refinement which is in GM(A,B),

(11.7.c) Let R denote the class of pairs $(\alpha,\beta) \in GM(A,B)^2$ such that α and β are similar. Then R is a set, and is an equivalence relation on GM(A,B).

The set of equivalence classes, under R of (11.7.c), is denoted by GC(A,B). A member of GC(A,B) is called a similarity, or a global, class from A to B. For $\alpha = (\theta,f)$ a global map from $A \longrightarrow B$, the member of GC(A,B) consisting of all $\beta \in GM(A,B)$ similar to α is denoted by $[\alpha]$ or $[\theta,f]$. Note that if S is a set of global maps from A to B, then $\alpha \mapsto [\alpha]$ is a set-theoretic function $S \longrightarrow GC(A,B)$.

Proof: Trivial. □

It is inconvenient to reduce every global map of interest to a representative in GM(*,*). Throughout the discussion, we work with general global maps, and leave it to the reader to transform elements into normalized form whenever rigor so demands. For $\alpha \in GC(A,B)$, a global map f is said to be of type α if $\alpha = [f]$.

Lemma 11.8: Suppose $A,B,C \in \mathcal{C}$

(A) Suppose (θ,f) is a global map from A to B and (φ,g) is a global map from B to C. There exists a cover η of A such that

(11.9.a) η factors through θ, and

(11.9.b) $(\theta,f)\circ\eta$ factors through φ.

(B) Suppose $A,B,C \in \mathcal{C}$, $\alpha \in GC(A,B)$ and $\beta \in GC(B,C)$. There is a unique $\gamma \in GC(A,C)$ such that if (Θ,f) is a global map of type α, (φ,g) is a global map of type β, and η is a cover of A which satisfies (11.9.a,b), then $(\eta,(\varphi,g)\circ\{(\Theta,f)\circ\eta\})$ is of type γ. We denote γ by $\beta\circ\alpha$, and call it the composition of β with α.

<u>Proof:</u> (A) An example is

(11.10) $$(y,z)\longmapsto \Theta(y)\circ f(y)^{-1}\varphi(z) \qquad \text{on } \mathrm{dom}(\Theta)\times\mathrm{dom}(\varphi).$$

(B) Suppose $(\Theta,f),(\Theta',f')$ are of type α, $(\varphi,g),(\varphi',g')$ are of type β, and η and η' are covers of A such that η satisfies (11.9.a,b) with respect to (Θ,f) and (φ,g), and η' meets the analogous conditions with respect to $(\Theta'.f')$ and (φ',g'). We must show that $(\eta,(\varphi,g)\circ\{(\Theta,f)\circ\eta\})$ and $(\eta',(\varphi',g')\circ\{(\Theta',f')\circ\eta\})$ are similar.

First, suppose Θ_0 is a refinement of Θ and Θ' such that $(\Theta,f)\circ\Theta_0=(\Theta',f')\circ\Theta_0$. Let η_0 be a cover of A which refines Θ_0, η and η'. Now

(11.11) $$(\eta,(\varphi,g)\circ\{(\Theta,f)\circ\eta\})\circ\eta_0=(\varphi,g)\circ(\{(\Theta,f)\circ\eta\}\circ\eta_0)$$
$$=(\varphi,g)\circ\{(\Theta,f)\circ\eta_0\}=(\varphi,g)\circ(\{(\Theta,f)\circ\Theta_0\}\circ\eta_0).$$

The analogue holds for the primed data. We are reduced to the case

(11.12) $$\Theta=\Theta',\quad f=f'\quad\text{and}\quad \eta=\eta'.$$

Hereafter, assume (11.12).

Let φ_0 be a refinement of φ and φ' such that $(\varphi,g)\circ\varphi_0=(\varphi',g')\circ\varphi_0$. Let η_0 be a cover of A which refines η and for which $(\Theta,f)\circ\eta_0$ factors through φ_0. Tracing through

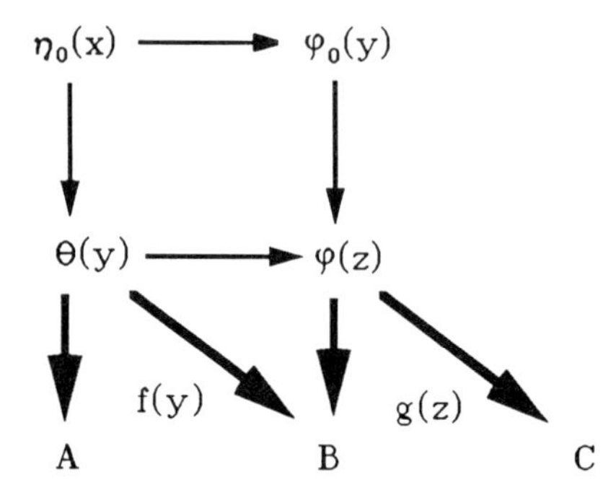

(11.13)

yields

(11.14) $$(\varphi_0,(\varphi,g)\circ\varphi_0)\circ\{(\Theta,f)\circ\eta_0\}=(\varphi,g)\circ\{(\Theta,f)\circ\eta_0\}=(\eta,(\varphi,g)\circ\{(\Theta,f)\circ\eta\})\circ\eta_0.$$

The analoguous equality holds for (φ',g'). Consequently, refinements through η_0 of the two compositions agree. $\square$

Definition 11.15: Let $\mathcal{O}$ be the object class of $\mathcal{C}$. For $(A,B,C) \in \mathcal{O}^3$, there is a function $c(A,B,C) : GC(B,C) \times GC(A,B) \longrightarrow GC(A,C)$ determined $(\beta,\alpha) \mapsto \beta \circ \alpha$. For $A \in \mathcal{O}$, let $\mathbb{1}_A$ be the similarity class of the global map $(\{(1,1_A)\},\{(1,1_A)\})$ from A to A. The following data determines a category:

(11.16.a) the objectclass $\mathcal{O}$,

(11.16.b) the morphism assignment $(A,B) \mapsto GC(A,B)$ on $\mathcal{O}^2$,

(11.16.c) the composition assignment $(A,B,C) \mapsto c(A,B,C)$ on $\mathcal{O}^3$,

(11.16.d) the identity assignment $A \mapsto \mathbb{1}_A$ on $\mathcal{O}$.

We refer to this category by $\mathcal{C}^s$, the smoothing of $\mathcal{C}$. We continue to write GC(A,B) for $\mathrm{Mor}_{\mathcal{C}^s}(A,B)$.

For $A,B \in \mathcal{C}$ and $b \in \mathrm{Mor}_{\mathcal{C}}(A,B)$, define

(11.17) $b^s = [\{(1,1_A)\},\{(1,b)\}] \in GC(A,B)$.

Then the class-theoretic assignments $A \mapsto A, b \mapsto b^s$ define a covariant functor $\mathcal{C} \longrightarrow \mathcal{C}^s$, which we call the smoothing functor.

Proposition 7.5 now gives us the useful tautology

Proposition 11.18: If $b : B \longrightarrow A$ is a monomorphic flush local subset of $\mathcal{C}$, then b^s is a $\mathcal{C}^s$-isomorphism.

Definition 11.19: Suppose (θ,f) is a global map from $A \longrightarrow B$.

Let $\theta^\# : A_0 \longrightarrow A^p$ be a canopy of θ. Then $f^\# : x \mapsto [f(x),1]$ determines a $\mathcal{C}^p$-morphism $A_0 \longrightarrow B^p$, which we call the formalization of (θ,f) through $\theta^\#$, or, sometimes, a formalization of f. The complete triple $(A_0,\theta^\#,f^\#)$ is also referred to as a formalization of (θ,f). Now assume, in addition, that φ is a cover of B through which f factors. A formalization $\varphi^\# : B_0 \longrightarrow B^p$ is monomorphic, and it is simple to produce the unique $f_0 : A_0 \longrightarrow B_0$ for which $\varphi^\# \circ f_0 = f^p \circ \theta^\#$. Such a morphism f_0 is called a formalization of (θ,f) through $\theta^\#$ and $\varphi^\#$, or through A_0 and B_0.

Smoothing is primarily of interest in the case $\mathcal{C} = \mathcal{D}^p$ for some topologized category $\mathcal{D}$ which meets the smallness condition. In this case, canopies in $\mathcal{C}$ have affinizations. It is natural to 'pushdown' formalizations from $\mathcal{D}^{pp}$ to $\mathcal{D}^p$.

Suppose $\mathfrak{C}$, A, B, (θ,f), A_0, $\theta^{\#}$ and $f^{\#}$ are as above, and now suppose $\theta_0 : A_0 \longrightarrow A'$ is an affinization with respect to $\mathfrak{C}$. Let $\theta' : A' \longrightarrow A$ so that $\theta^{\#} = \theta' \circ \theta_0$. Tautologically, θ' admits an inverse in $\mathfrak{C}^S$. The factoring $f' : A' \longrightarrow B$ of $f^{\#}$ (regarded as a cone) is called a pushdown formalization with respect to the data. This term is also applied to the triple (A_0,θ_0,f'). Moreover, θ' yields a $\mathfrak{C}^S/B^S$-isomorphism between $(A,[\theta,f])$ and (A',f'^S). Again, if φ, $\varphi^{\#} : B_0 \longrightarrow B$ and f_0 are as above, and $\varphi' : B_0 \longrightarrow B'$ is a colimit, there is a unique $f'' : A' \longrightarrow B'$ which factors the cone of the $\mathfrak{C}^P$-morphism $\varphi \circ f_0$. We also refer to f'' as a pushdown formalization.

Let (S,M) be a graph. We say (S,M) (or any graph of type (S,M)) is finite if S and M are finite sets. Our immediate aims are (a) to show that the smoothing functor preserves finite inverse limits, and (b) to find a criterion for existence of fibered products.

Proposition 11.20: Let (S,M) be a finite graph. Let G be a graph of $\mathfrak{C}^P$-objects of type (S,M), and suppose $\theta : A_0 \longrightarrow G$ is a source in $\mathfrak{C}^P$. For each $x \in S$, let $B(x) \in \mathfrak{C}$, and suppose $x^{\#} : G(x) \longrightarrow B(x)$ is the canopy of a cover. Suppose $A \in \mathfrak{C}$ and $\alpha^{\#} : A_0 \longrightarrow A$ is the canopy of a cover.

Define a graph G' in $\mathfrak{C}^S$ by

(11.21) $\qquad G'(x) = B(x) \qquad\qquad$ for $x \in S$,

and, for $(x,\rho,y) \in M$, $G'(x,\rho,y)$ is the unique global class of which

(11.22) $\qquad (G(x), x^{\#}, y^{\#} \circ G(x,\rho,y))$

is a formalization. Define a source $\theta' : A \longrightarrow G$ by mapping each $x \in S$ to the unique global class of which

(11.23) $\qquad (A_0, \alpha^{\#}, x^{\#} \circ \theta(x))$

is a formalization.

(A) If θ is a pseudoinjective source of $\mathfrak{C}^P$, then θ' is a pseudoinjective source of $\mathfrak{C}^S$.

(B) If θ is an inverse limit in $\mathfrak{C}^P$, then θ' is an inverse limit in $\mathfrak{C}^S$.

Proof: Clearly G' is a graph in $\mathfrak{C}^S$, and θ' is a source. For $x \in S$, let x^* denote the cover which $x^{\#}$ formalizes. Let α^* denote the cover which $\alpha^{\#}$ formalizes.

(A) Suppose $D \in \mathfrak{C}$ and $f_1, g_1 \in GC(D,A)$ such that $\theta'(x) \circ f_1 = \theta'(x) \circ g_1$ for each $x \in S$. There is a cover δ of D and global maps (δ,f) and (δ,g) of type f_1 and g_1, respectively, such that f and g each factor through α.

For each $x \in S$, there is a refinement δ' of δ such that

(11.24) $\{(\alpha^{\#}, x^{\#} \circ \Theta(x)) \circ (\delta, f)\} \circ \delta' = \{(\alpha^{\#}, x^{\#} \circ \Theta(x)) \circ (\delta, g)\} \circ \delta'$.

As S is finite, passing to a common refinement and invoking (11.5) reduces the problem to the case in which

(11.25) $(\alpha^{\#}, x^{\#} \circ \Theta(x)) \circ (\delta, f) = (\alpha^{\#}, x^{\#} \circ \Theta(x)) \circ (\delta, g)$

for each $x \in S$.

Let f_0 and g_0 be formalizations of f and g, respectively, through $\delta^{\#}$ and $\alpha^{\#}$. For $x \in S$, $x^{\#}$ is monomorphic, from which we conclude $\Theta(x) \circ f_0 = \Theta(x) \circ g_0$. Since Θ is pseudoinjective, we get

(11.26) $f_0 = g_0 \;\Rightarrow\; f = g \;\Rightarrow\; f_1 = g_1$.

(B) Suppose $\varphi : D \longrightarrow G'$ is a source in $\mathcal{C}^S$. For each $x \in S$, let (δ_x, f_x) be a representative of $\varphi(x)$. We may pass to a cover δ of D which refines δ_x for each $x \in S$; assume $\delta = \delta_x$ for each $x \in S$. For each $x \in S$, there is a refinement δ' of δ such that $(\delta, f_x) \circ \delta'$ factors through $x^{\#}$. As S is finite, we may assume (δ, f_x) factors through $x^{\#}$ for every $x \in S$. For $(x, \rho, y) \in M$, there is a refinement δ' of δ such that

(11.27) $\{(x^{*}, y^{\#} \circ G(x, \rho, y)) \circ (\delta, f_x)\} \circ \delta' = (x^{*}, y^{\#} \circ G(x, y, \rho)) \circ \{(\delta, f_x) \circ \delta'\}$

and $(\delta, f_y) \circ \delta'$ agree. As M is finite, after passing to a suitable common refinement we may assume

(11.28) $(x^{*}, y^{\#} \circ G(x, \rho, y)) \circ (\delta, f_x) = f_y$

for each $(x, \rho, y) \in M$.

Fix a canopy $\delta^{\#} : D_0 \longrightarrow D$ of δ. For each $x \in S$, let f^x be the formalization of f_x through $\delta^{\#}$ and $x^{\#}$. Since canopy morphisms are monomorphic, (11.28) yields that $x \mapsto f^x$ on S is a source of G. Fix $f_0 : D_0 \longrightarrow A_0$ such that $\Theta(x) \circ f_0 = f^x$ for each $x \in S$. Then $\Theta'(x) \circ (\delta, \alpha^{\#} \circ f_0) = \varphi(x)$ for each $x \in S$. □

Corollary 11.29: If $b : B \longrightarrow C$ is a $\mathcal{C}$-monomorphism, then b^S is monomorphic.

Corollary 11.30: Let G be a finite graph of $\mathcal{C}$-objects, and let $\Theta : A \longrightarrow G$ be an inverse limit. Then the image of Θ under smoothing is an inverse limit.

Theorem 11.31: Suppose $B, X_1, X_2 \in \mathcal{C}$, and suppose (Θ_j, f_j) is a global map from X_j to B for each $j \in \{1,2\}$. For $j \in \{1,2\}$, let $\theta_j^{\#}: Y_j \longrightarrow X_j$ be a canopy of θ_j, and let $f_j^{\#}$ be the formalization of (θ_j, f_j) through Y_j. Suppose

(11.32.a) $(P;\pi_1,\pi_2)$ is a fibered product $(Y_1,f_1^{\#})\times_{B^P}(Y_2,f_2^{\#})$ in $\mathcal{C}^P$,

(11.32.b) there is a $\mathcal{C}$-cover φ of some $C \in \mathcal{C}$ of which P is a canopy.

For $j \in \{1,2\}$, define q_j a function on $\Lambda(P)$ by the condition that for each $x \in \Lambda(P)$, $q_j(x) \in \mathrm{Mor}_{\mathcal{C}}(P(x),X_j)$ and $q_j^P = \theta_j^{\#} \circ \pi_j \circ \iota_{P;x}$ (where $\iota_{P;x}$ is the x-th coordinate chart of P). Then $(C;[\varphi,q_1],[\varphi,q_2])$ is a fibered product of $(X_1,[\theta_1,f_1])\times_B(X_2,[\theta_2,f_2])$ in the category $\mathcal{C}^S$.

Corollary 11.33: Suppose $\mathcal{D}$ is a topologized category and $\mathcal{C} = \mathcal{D}^P$. Suppose (θ,f) is a global map between two $\mathcal{C}$-objects. If $f(x)$ is a pullback base for each $x \in \mathrm{dom}(\theta)$, then $[\theta,f]$ is a pullback base of $\mathcal{C}^S$.

We often use the hypothesis

(12.1) $(\mathcal{C},\mathrm{Sub},\mathrm{Cov})$ is a category with a Grothendieck topology which satisfies the smallness condition, and $\mathcal{C}\dashrightarrow\mathcal{C}^s$ is a choice of smoothing functor of $\mathcal{C}$.

Let $(\theta,f):B\longrightarrow A$ be a global map. The following are tautological:

(12.2.a) If $f':B'\longrightarrow A$ is a formalization of (θ,f), then (B',f'^s) and $(B,[\theta,f])$ are $\mathcal{C}^s/A^s$-isomorphic.

(12.2.b) For each $j\in\mathrm{dom}(\theta)$, $f(j)^s=[\theta,f]\circ\theta(j)^s$.

We call (θ,f) an explicit local subset if $f(j)\in\mathrm{Sub}$ for each $j\in\mathrm{dom}(\theta)$. A global class is called an e.l.-class if it contains an explicit local subset.

By a choice of plus functor for $\mathcal{C}$, we mean a functor $F:\mathcal{C}\dashrightarrow\mathcal{D}$ where $\mathcal{D}$ is the smoothing of $\mathcal{C}^p$ with respect to some categorical choice of representatives and F is the composition of pasting and smoothing functors. When dealing with a plus functor, the symbols 'p' and 's' denote the intermediate constructions. We refer to $\mathcal{D}$ by $\mathcal{C}^+$ and to F as +, and write α^+ to denote the image of object or morphism under +.

Suppose b is a formal $\mathcal{C}$-subset. Theorem 11.31 assures that b^s is a pullback base provided that $\mathcal{C}$ has enough affinizations. For our purposes, the smoothing operation is useful only in conjunction with the pasting functor—that is, when affinizations abound. Some technicalities can be phrased in terms of simple smoothing.

Lemma 12.3: Assume (12.1). Let $(\theta,f):B\longrightarrow A$ be a global map.

(A) $[\theta,f]$ is an e.l.-class if and only if the formalization of (θ,f) is a local subset in $\mathcal{C}^p$.

(B) Suppose a pushdown formalization f' of (θ,f) exists. Then $[\theta,f]$ is an e.l.-class if and only if f' is a local subset of $\mathcal{C}$.

Proof: (A) Let $f^{\#}:B_0\longrightarrow A^p$ be a formalization. Direct application of the definition of Cov^p implies that if $f^{\#}$ is a local subset, then $[\theta,f]$ is an e.l.-class. Conversely, suppose η is a refinement of θ such that $(\theta,f)\circ\eta$ assumes only subsets. It suffices to show that for $x\in\mathrm{dom}(\eta)$ and $y\in\mathrm{dom}(\theta)$, $f^{\#}\circ\iota_y\circ\pi_y=f(y)^p\circ\pi_y$ on $\eta(x)\times_A\theta(y)$ is a formal subset. Fix $(x,y)\in\mathrm{dom}(\eta)\times\mathrm{dom}(\theta)$, and take $z\in\mathrm{dom}(\theta)$ and $\gamma\in\mathrm{Mor}_{\mathcal{C}/A}(\eta(x),\theta(z))$. Factoring through $\theta(z)\times_A\theta(y)$ yields

(12.4) $$f(y)^p \circ \pi_y = \{f(y) \circ \pi_y \circ (\gamma \times 1_{\Theta(y)})\}^p = \{f(z) \circ \pi_z \circ (\gamma \times 1_{\Theta(y)})\}^p$$
$$= \{f(z) \circ \gamma \circ \pi_x\}^p.$$

Both $f(z) \circ \gamma$ and π_x are formal subsets.

(B) is a trivial consequence of (A). □

Definition 12.5: Let $\mathcal{D}$ be a topologized category. We say $\mathcal{D}$ is quasi-intrinsic if for $A,B \in \mathcal{D}$, Θ an indexed cover of B, and $f,g \in \mathrm{Mor}_{\mathcal{C}}(B,A)$ such that $f \circ \Theta(j) = g \circ \Theta(j)$ for all $j \in \mathrm{dom}(\Theta)$, it must follow that $f = g$.

Proposition 12.6: Assume (12.1). Suppose $A,B \in \mathcal{C}$ and Θ is a $\mathcal{C}$-cover of B.

(A) Suppose $f_1, g_1 \in \mathrm{Mor}_{\mathcal{C}^s}(B,A)$ so that $f_1 \circ \Theta(j)^s = g_1 \circ \Theta(j)^s$ for each $j \in \mathrm{dom}(\Theta)$. Then $f_1 = g_1$.

(B) Suppose the topology of $\mathcal{C}$ is quasi-intrinsic. The function $f \mapsto f^s$ from $\mathrm{Mor}_{\mathcal{C}}(B,A) \longrightarrow GC(B,A)$ is injective.

(C) Suppose the topology of $\mathcal{C}$ is quasi-intrinsic. Let $\Theta^{\#} : B_0 \longrightarrow B$ be a canopy of Θ. Then the image of $\Theta^{\#}$ under the smoothing functor is a colimit.

Proof: (A) Let (σ,f) and (τ,g) represent f_1 and g_1, respectively. Passing to a mutual refinement, we may assume that $\sigma = \tau = \Theta$. For each $j \in \mathrm{dom}(\Theta)$, there is a cover φ_j of $\mathrm{dom}\,\Theta(j)$ such that $f(j) \circ \varphi_j(x) = g(j) \circ \varphi_j(x)$ for each $x \in \mathrm{dom}(\varphi_j)$. Clearly the refinements of (Θ,f) and (Θ,g) through the subdivision of $j \mapsto \varphi_j$ agree.

(B) is a tautology.

(C) Suppose $\beta : B_0 \longrightarrow A$ is a cone with respect to $\mathcal{C}^s$. For each $j \in \mathrm{dom}(\Theta)$, let (φ_j, f_j) represent $\beta(j)$. Let φ denote the subdivision of Θ by $j \mapsto \varphi_j$, and define f on $\Lambda = \mathrm{dom}(\varphi)$ by $f(j,x) = f_j(x)$.

Suppose $(j,x),(k,y) \in \Lambda$. On $\varphi(j,x) \times_B \varphi(k,y)$,

(12.7) $$\{f(j,x) \circ \pi_{j,x}\}^s = \beta(j) \circ \{\varphi_j(x) \circ \pi_{j,x}\}^s$$
$$= \beta(j) \circ \{\pi_j \circ (\varphi_j(x) \times \varphi_k(y))\}^s = \beta(j) \circ \pi_j^{\,s} \circ \{(\varphi_j(x) \times \varphi_k(y))\}^s.$$

In the same manner,

(12.8) $$\{f(k,y) \circ \pi_{k,y}\}^s = \beta(k) \circ \pi_k^{\,s} \circ \{\varphi_j(x) \times \varphi_k(y)\}^s.$$

By hypothesis, $\beta(j)\circ\pi_j{}^s = \beta(k)\circ\pi_k{}^s$. Citing (B), we deduce that $f(j,x)\circ\pi_{j,x} = f(k,y)\circ\pi_{k,y}$. Consequently, (φ,f) is a global map $A \longrightarrow B$.

Let $j \in \mathrm{dom}(\Theta)$. The pair

(12.9) $\quad ((k,y)\mapsto(\Theta(j)\times_A\varphi(k,y),\pi_j),(k,y)\mapsto f(k,y)\circ\pi_{k,y})$

represent $[\varphi,f]\circ\Theta(j)^s$. The preceeding paragraph states that the two refinements of (φ_j,f_j) and (12.9) through $(x,(k,y))\mapsto\varphi(j,x)\times_A\varphi(k,y)$, on $\mathrm{dom}(\varphi_j)\times\Lambda$, agree. Thus, $[\varphi,f]\circ\Theta(j)^s = \beta(j)$. □

Lemma 12.10: Assume (12.1). If $\mathcal{C}$ is intrinsic, then $\mathcal{C}^p$ is quasi-intrinsic.

Proof: The theorem easily reduces to the following statement:

(12.11) Suppose $A \in \mathcal{C}$, Θ is a $\mathcal{C}$-cover of A, $B_0 \in \mathcal{C}^p$ and $f_0,g_0 \in \mathrm{Mor}_{\mathcal{C}^p}(A^p,B_0)$ so that $f_0\circ\Theta(x)^p = g_0\circ\Theta(x)^p$ for each $x \in \mathrm{dom}(\Theta)$. Then $f_0 = g_0$.

Assume the hypothesis of (12.11). Let (f,j) and (g,k) represent f_0 and g_0, respectively.

For each $x \in \mathrm{dom}(\Theta)$, take $h(x) \in \mathrm{Mor}_{\mathcal{C}}(A,B_0(j,k))$ so that $\rho_j\circ h(x) = f\circ\Theta(x)$ and $\rho_k\circ h(x) = g\circ\Theta(x)$. From (3.5.b), it follows that $x\mapsto h(x)$ extends to a cone on the canopy of Θ. As $\mathcal{C}$ is intrinsic, there is $h \in \mathrm{Mor}_{\mathcal{C}}(A,B_0(j,k))$ so that $h\circ\Theta(x) = h(x)$ for each $x \in \mathrm{dom}(\Theta)$. Trivially, $\rho_j\circ h = f$ and $\rho_k\circ h = g$. □

We would like to phrase the preceeding results in terms of Grothendieck topologies. The first obstruction is to lift the notion of a formal subset to $\mathcal{C}^s$—roughly, to recognize when a $\mathcal{C}^s$-morphism is isomorphic to a subset from $\mathcal{C}$. A characterization of $\mathcal{C}^s$-isomorphisms is necessary.

Definition 12.12: Let $\mathcal{D}$ be a topologized category. A $\mathcal{D}$-morphism $b: B \longrightarrow A$ is called a pseudoisomorphism if there is a cover α of A such that for each $j \in \mathrm{dom}(\alpha)$, $\alpha(j)^{-1}b$ is a $\mathcal{D}$-isomorphism.

Remark 12.13: Assume (12.1). Let $(\Theta,f): B \longrightarrow A$ be a global map. Then $(\Theta,f)^p$, the pair of compositions of Θ and f with the pasting functor, is a global map $B^p \longrightarrow A^p$. It is easy to check that (Θ,f) is a $\mathfrak{C}^s$-isomorphism if and only if $(\Theta,f)^p$ is a $\mathfrak{C}^{ps}$-isomorphism. Now suppose $f^{\#}$ is a formalization of (Θ,f). Then $f^{\#}$ can be regarded as a pushdown formalization in $\mathfrak{C}^p$, and so is equivalent to $[(\Theta,f)^p]$ under smoothing!

Lemma 12.14: Assume (12.1). Let $f \in \mathrm{Mor}(\mathfrak{C})$.

(A) If f is a pseudoisomorphism, then f^s is a $\mathfrak{C}^s$-isomorphism.

(B) Suppose $\mathfrak{C}$ is quasi-intrinsic and f^s is a $\mathfrak{C}^s$-isomorphism. Then f is a pseudoisomorphism. In particular, f is a local subset.

Proof: (A) is tautological.

(B) Assume $\mathfrak{C}$ is quasi-intrinsic and f^s is a $\mathfrak{C}^s$-isomorphism. Suppose (φ,g) represents an inverse to f^s. After passing to a refinement, we may assume that $(\{(1,1_A)\},f)\circ g = \varphi$. Let $j \in \mathrm{dom}(\varphi)$. Theorem 11.31 suggests that the pullback of f along $\varphi(j)$ also represents a $\mathfrak{C}^s$-isomorphism. To finish (B), it suffices to prove that each of these pullbacks of f is a pseudoisomorphism. The problem reduces to the case in which there exists $g \in \mathrm{Mor}_{\mathfrak{C}}(A,B)$ so that $f\circ g = 1_A$.

Obviously $g^s = (f^s)^{-1}$. Thus, there is a cover η on A such that for each $x \in \mathrm{dom}(\eta)$, $g\circ f\circ \eta(x) = \eta(x)$. But $\mathfrak{C}$ is quasi-intrinsic, so $g\circ f = 1_B$. □

Corollary 12.15: Let $\mathfrak{C}$ be a category with an intrinsic Grothendieck topology which meets the smallness condition. Fix a choice of plus functor for $\mathfrak{C}$. Let (Θ,f) be a global map with respect to $\mathfrak{C}^p$. Then the following conditions on (Θ,f) are equivalent:

(12.16.a) $[\Theta,f]$ is a $\mathfrak{C}^+$-isomorphism.

(12.16.b) A pushdown formalization of (Θ,f) is a pseudoisomorphism in $\mathfrak{C}^p$.

The main result of this section is now a simple exercise.

Theorem 12.17: Assume (12.1). Suppose $\mathfrak{C}$ is complete under affinization and that its topology is quasi-intrinsic. Let Sub_1 denote the class of e.l.-classes in $\mathfrak{C}^s$.

Let Θ be a non-empty indexed cone in $\mathfrak{C}^s$. Let Θ_1 be a function which to each

$j \in \mathrm{dom}(\Theta)$ assigns $(\varphi_j, g_j) \in \Theta(j)$. Let

(12.18) $\Lambda = \{(j,x) : j \in \mathrm{dom}(\Theta),\ x \in \mathrm{dom}(\varphi_j)\}$.

Suppose the function $(j,x) \mapsto g_j(x)$ is a cover of $\mathrm{cod}\,\Theta$ in the topology of $\mathcal{C}$. In this case, we say Θ is an e.l.-cover of $\mathrm{cod}\,\Theta$. Let Cov_1 be the class of all sets of $\mathcal{C}^S$-morphisms which are e.l.-covers.

(A) Sub_1 is a universe of subsets for $\mathcal{C}^S$, and Cov_1 is a topology defined with respect to Sub_1. We refer to $(\mathrm{Sub}_1, \mathrm{Cov}_1)$ as the e.l.-topology.

(B) Cov_1 is intrinsic.

(C) If Cov is flush, then so is Cov_1.

(D) If Sub is a universe of embeddings, then so is Sub_1.

Proof: (A) Any pushdown formalization of a $\mathcal{C}^S$-isomorphism is known to be a pseudoisomorphism. Thus, every $\mathcal{C}^S$-isomorphism is an e.l.-class. By Corollary 10.4, each pushdown formalization of an e.l.-class is a pullback base in $\mathcal{C}$. It is now trivial verify that Sub_1 is a universe of subsets.

It is simple to check that Cov_1 meets (2.9.a,b,c,d). The last two conditions involve products of $\mathcal{C}^S$-morphisms; unfortunately, the characterization of Theorem 11.31 is unweildy.

Suppose $X \in \mathcal{C}$ and Θ is a non-empty indexed subsets of $\mathcal{C}^S/X^S$. Suppose φ assigns to each $j \in \mathrm{dom}(\Theta)$ a $\mathcal{C}^+$-isomorphism into $\mathrm{dom}\,\Theta(j)$, and $\tau : X \longrightarrow Y$ is a $\mathcal{C}^+$-isomorphism. Define Θ_1 and Θ_2 on $\mathrm{dom}(\Theta)$ by $\Theta_1(j) = \Theta(j) \circ \varphi(j)$ and $\Theta_2(j) = \tau \circ \Theta(j)$. Properties (2.9.c,d) imply that the conditions

(12.19.a) Θ is an e.l.-cover,

(12.19.b) Θ_1 is an e.l.-cover,

(12.19.c) Θ_2 is an e.l.-cover,

are equivalent! Now for $Z \in \mathcal{C}$, each object in $\mathcal{C}^+/Z^+$ is $\mathcal{C}^+/Z^+$-isomorphic to (B^S, b^S) where $(B,b) \in \mathcal{C}/Z$. The above equivalence reduces proof of (2.9.e,f) to the cases in which all morphisms are images of $\mathcal{C}$-morphisms. Verification is now trivial.

(B) We must show that every set in Cov_1 is an intrinsic cover. It suffices to show that if S is a non-empty cone of $\mathcal{C}$ whose image under smoothing is in Cov_1, then S is intrinsic. Proposition 2.24 reduces this to showing that smoothing maps $\mathcal{C}$-covers to absolute covers.

Let Θ be a $\mathcal{C}$-cover, and let Θ^S denote its image under smoothing. Let $\alpha : G \longrightarrow \mathrm{cod}\,\Theta$ be a canopy of Θ, and let α^S denote its image under smoothing. Each pullback of α^S along a $\mathcal{C}^S$-morphism is isomorphic to a pullback of α along a $\mathcal{C}$-morphism under smoothing. Hence, to show that Θ is absolute, it suffices to show that every $\mathcal{C}$-cover is mapped to an intrinsic cover. This is Proposition 12.6.

(C) Identifying each $\mathcal{C}^S$-morphism with a $\mathcal{C}$-morphism reduces the question to Proposition 2.16.

(D) follows from Proposition 2.14. □

Remark 12.20: Let $\mathcal{C}$ be a category with an intrinsic Grothendieck topology which meets the smallness condition. Fix a plus functor for $\mathcal{C}$. The e.l.-topology on $\mathcal{C}^+$ is too large. The objection is that every local subset of $\mathcal{C}$ is a formal e.l.-subset. Still, the criterion for determining which sets of e.l.-classes form a cover is correct. Ultimately, Proposition 2.36 enables us to shrink the notion of subset.

Pseudoisomorphisms were introduced as a technical tool. In Part V, they offer a generalization of affinizations. Some closing remarks will be useful later.

Lemma 12.21: Let $\mathcal{C}$ be a quasi-intrinsic topologized category. Let $b: B \longrightarrow A$ be a $\mathcal{C}$-morphism and let Θ be an indexed cover of A such that $\Theta(j)^{-1}b$ is monomorphic for each $j \in \mathrm{dom}(\Theta)$. Then b is monomorphic.

Proof: Suppose $C \in \mathcal{C}$ and $f,g \in \mathrm{Mor}_{\mathcal{C}}(C,A)$ so that $b \circ f = b \circ g$. We claim that $f = g$.

Put $\gamma = b \circ f$. For $j \in \mathrm{dom}(\Theta)$, let $(P(j);\pi_{j,C},\pi_j)$ be a fibered product $(C,\gamma) \times_A \Theta(j)$. Hypothesis reduces the problem to showing that $f \circ \pi_{j,C} = g \circ \pi_{j,C}$ for each j. Hereafter, fix $j \in \mathrm{dom}(\Theta)$.

Let $(P;\pi_B,\pi')$ be a fibered product $b \times_A \Theta(j)$. Let f_1 and g_1 be the respective morphisms $f \times 1_{\Theta(j)}$ and $g \times 1_{\Theta(j)}$ from $P(j)$ to P. Then π' is monomorphic and

(12.22) $$\pi' \circ f_1 = \pi_j = \pi' \circ g_1 \quad \Rightarrow \quad f_1 = g_1$$

$$\Rightarrow \quad f \circ \pi_{j,C} = \pi_B \circ f_1 = \pi_B \circ g_1 = g \circ \pi_{j,C}. \quad \square$$

Proposition 12.23: Let $\mathfrak{C}$ be a quasi-intrinsic topologized category. A $\mathfrak{C}$-morphism is a pseudoisomorphism if and only if it is a monomorphic flush local subset.

Proof: Let $b \in \mathrm{Mor}(\mathfrak{C})$. If b is a pseudoisomorphism, Lemma 12.21 implies that it is a monomorphic flush local subset. Conversely, suppose b is a monomorphism and S is a cover of dom b such that $\{b \circ s : s \in S\}$ is a cover of cod b. Proposition 7.1 easily implies that pullback of b along $b \circ s$ is an isomorphism for each $s \in S$.

Proposition 12.24: Let $\mathfrak{C}$ be a topologized category. Assume the Grothendieck topology of $\mathfrak{C}$ is intrinsic, flush, and satisfies the CLCS and smallness conditions.

(A) The class of pseudoisomorphisms in $\mathfrak{C}^P$ forms a universe of embeddings.

(B) Let $A \in \mathfrak{C}$ and let $b_0 : B_0 \longrightarrow A^P$ be a $\mathfrak{C}^P$-morphism. Then b_0 is a pseudoisomorphism if and only if b_0 is an affinization.

Proof: Previous lemmas will yield (A) once we prove that a pseudoisomorphism of $\mathfrak{C}^P$ is a pullback base. To show $b_0 \in \mathrm{Mor}(\mathfrak{C}^P)$ is a pullback base, it suffices to show that each of its pullbacks along members of the assigned cover is a pullback base; when b_0 is a pseudoisomorphism, so is each pullback. We are reduced to showing (B). Proposition 12.23 states that affinizations are pseudoisomorphisms.

Suppose $A \in \mathfrak{C}$, $B_0 \in \mathfrak{C}^P$ and $b_0 \in \mathrm{Mor}_{\mathfrak{C}}(B_0, A^P)$ is a pseudoisomorphism. We claim b_0 is an affinization. Previous lemmas and the assumption that $\mathfrak{C}$ is flush reduce the claim to showing that $b_0 \circ \iota_j$ is a formal subset for each $j \in \Lambda(B_0)$. Fix $j \in \Lambda(B_0)$ and let $f \in \mathrm{Mor}_{\mathfrak{C}}(B_0(j), A)$ so $f^P = b_0 \circ \iota_j$. From Proposition 12.23, f is a local $\mathfrak{C}$-subset and $(B_0(j,j)^P; \rho_1{}^P, \rho_2{}^P)$ serves as a fibered product $f^P \times_{A^P} f^P$. The CLCS condition implies that f is a formal subset. □

§13 Functorial Properties and Smoothing

Assume (12.1) and denote the smoothing functor by s.

Theorem 13.1: Let $\mathcal{D}$ be a category.

(A) Suppose $\Gamma : \mathcal{C} \longrightarrow \mathcal{D}$ is a functor of sections. Then there is a unique functor $\Phi : \mathcal{C}^s \longrightarrow \mathcal{D}$ such that $\Phi \circ s = \Gamma$.

(B) Suppose the topology of $\mathcal{C}$ is quasi-intrinsic and $\mathcal{C}$ is closed under affinization. Suppose $\Phi : \mathcal{C}^s \longrightarrow \mathcal{D}$ is a covariant functor for which $\Phi \circ s$ is a functor of sections. Then Φ is a functor of sections with respect to the e.l.-topology.

(C) Suppose the topology of $\mathcal{C}$ is quasi-intrinsic and complete under affinization. Let (Sub',Cov') be a flush topology on $\mathcal{D}$. Suppose $\Phi : \mathcal{C}^s \longrightarrow \mathcal{D}$ is a covariant functor such that $\Phi \circ s$ is continuous. If Θ is an indexed e.l.-cover in $\mathcal{C}^s$ for which $\Phi(\Theta(j)) \in$ Sub' for each $j \in \mathrm{dom}(\Theta)$, then Φ maps Θ to an indexed cover with respect to Cov'.

Proof: (A) is tautological.

(B) Let $A \in \mathcal{C}$ and let Θ_1 be an indexed e.l.-cover of A^s. Let $\alpha_1 : A_1 \longrightarrow A^s$ be a canopy of Θ_1. We must show that $\Phi(\alpha_1)$ is a colimit. Without loss of generality, we may assume Θ_1 is a family Θ of $\mathcal{C}/A$ under smoothing, and then α_1 is the image of a canopy $\alpha_0 : A_0 \longrightarrow A$. For each $j \in \mathrm{dom}(\Theta)$, let φ_j be a $\mathcal{C}$-cover of $\mathrm{dom}\,\Theta(j)$ such that $\Theta(j) \circ \varphi_j(x) \in$ Sub for each $x \in \mathrm{dom}(\varphi_j)$. Note that Θ is an unrefined cover in $\mathcal{C}$.

Let $D \in \mathcal{D}$ and let $\delta : \Phi(A_1) \longrightarrow D$ be a cone. To prove (B), it suffices to find $d \in \mathrm{Mor}_{\mathcal{D}}(\Phi(A), D)$ so that $d \circ \Phi(\{\Theta(j) \circ \varphi_j(x)\}^s) = \delta(j) \circ \Phi(\varphi_j(x)^s)$ for each $j \in \mathrm{dom}(\Theta)$ and $x \in \mathrm{dom}(\varphi_j)$. Hypothesis reduces the problem to checking that for each $j,k \in \mathrm{dom}(\Theta)$, $x \in \mathrm{dom}(\varphi_j)$ and $y \in \mathrm{dom}(\varphi_k)$,

$$\delta(j) \circ \Phi(\{\varphi_j(x) \circ \pi_x\}^s) = \delta(k) \circ \Phi(\{\varphi_k(y) \circ \pi_y\}^s) \quad \text{on } \Phi(\Theta(j) \circ \varphi_j(x)\} \times_A \{\Theta(k) \circ \varphi_k(y)). \tag{13.2}$$

But, for such indices, $\varphi_j(x) \circ \pi_x$ and $\varphi_k(y) \circ \pi_y$ factor through the projections of $\Theta(j) \times_A \Theta(k)$. Equality follows from choice of δ.

(C) is a tautology. □

Proposition 13.3: Let $\mathcal{D}$ be a topologized category. Assume $\mathcal{C}$ is quasi-intrinsic and that both $\mathcal{C}$ and $\mathcal{D}$ are complete under affinization. Let $\Phi : \mathcal{C}^s \longrightarrow \mathcal{D}$ such that $\Phi \circ s$ is continuous. Let $A, B \in \mathcal{C}$ and $(\Theta, f) \in GM(B, A)$. Suppose that for each $j \in dom(\Theta)$, $\Phi(f(j)^s)$ is a pullback base in $\mathcal{D}$. Then

(13.4.a) $\Phi([\Theta, f])$ is a pullback base in $\mathcal{D}$,

(13.4.b) If $(C, c) \in \mathcal{C}^s/A$, then Φ preserves any fibered product of $(C, c) \times_A (B, [\Theta, f])$.

Proof: From Corollary 10.4. □

PART V LOCAL AND GLOBAL STRUCTURES

§14 The Local and Global Axioms

In this section, we finally define the notions of local and global structure, and state the main theorem on the plus construction. Part of this result is existence of globalizations, under special hypothesis. The proof occupies the remainder of the paper.

The reader should look at the lengthy Remarks of this section. Their point is that certain concepts, such as base spaces, Hausdorff/separation and closed embeddings, integrate into the universal theory without difficulty.

Definition 14.1: Let ($\mathfrak{C}$,Sub,Cov) be a topologized category. Let Cvm denote the class of covering morphisms in Mor($\mathfrak{C}$). Suppose

(14.2.a) (Sub,Cov) meets the smallness condition,

(14.2.b) (Sub,Cov) is flush,

(14.2.c) (Sub,Cov) is intrinsic,

(14.2.d) (Sub,Cov) meets the CLCS condition,

(14.2.e) (Sub,Cov) is complete under Cvm.

In this case, $\mathfrak{C}$ (or its Grothendieck topology) is called a local structure.

Suppose $\mathfrak{C}$ meets (14.2.a,b,c,d). We remark that

(14.3) In classical cases, Cvm is often the class of $\mathfrak{C}$-isomorphisms. However, this is *not* a requirement of a local structure.

We call $\mathfrak{C}$ a global structure if it is complete under affinization. A globalization of $\mathfrak{C}$ is a continuous functor $\Phi : \mathfrak{C} \longrightarrow \mathcal{D}$ where $\mathcal{D}$ is a global structure such that every continuous functor into a global structure is functorially equivalent to a functor which factors through Φ.

The main results of the paper can now be stated.

Theorem 14.4: Let ($\mathfrak{C}$,Sub,Cov) be a local structure. Let $+ : \mathfrak{C} \longrightarrow \mathfrak{C}^+$ be a choice of plus functor. Recall that + preserves inverse limits of finite graphs.

There is a unique topology (Sub^+,Cov^+) on $\mathfrak{C}^+$ such that

(14.5.b) (Sub^+,Cov^+) is a local structure with respect to which smoothing $\mathfrak{C}^S \longrightarrow \mathfrak{C}^+$ is continuous,

(14.5.c) if (Sub',Cov') is another local structure on $\mathfrak{C}$ with respect to which smoothing is continuous, then $\mathrm{Sub}^+ \subseteq \mathrm{Sub}'$ and $\mathrm{Cov}^+ \subseteq \mathrm{Cov}'$.

We refer to $(\mathrm{Sub}^+, \mathrm{Cov}^+)$ as the <u>plus topology</u>, and regard $\mathcal{C}^+$ as a local structure in this way.

(A) If Sub is a universe of embeddings, then so is Sub^+. If every covering morphisms of $\mathcal{C}$ is a $\mathcal{C}$-isomorphism, then every covering morphisms of $\mathcal{C}^+$ is a $\mathcal{C}^+$-isomorphism.

(B) The plus functor is a CLCS functorial embedding.

(C) Suppose $\mathcal{D}$ is a category in which every intersection graph admits a colimit. Let $\Gamma : \mathcal{C} \longrightarrow \mathcal{D}$ be a functor of sections. Then there is a functor of sections $\Gamma^+ : \mathcal{C}^+ \longrightarrow \mathcal{D}$ such that $\Gamma^+ \circ +$ is functorially equivalent to Γ; moreover, any two functors of sections with the latter property are functorially equivalent.

(D) Suppose $\mathcal{D}$ is a global structure and $\Gamma : \mathcal{C} \longrightarrow \mathcal{D}$ is continuous. There is a continuous functor $\Gamma^+ : \mathcal{C}^+ \longrightarrow \mathcal{D}$ such that $\Gamma^+ \circ +$ is functorially equivalent to Γ; moreover, any two continuous functors with the latter property are functorially equivalent. Also, if $\Gamma^+ : \mathcal{C}^+ \longrightarrow \mathcal{D}$ is continuous, then Γ^+ is a CLCS functorial embedding if and only if $\Gamma^+ \circ +$ is a CLCS functorial embedding.

(E) Let $+' : \mathcal{C}^+ \longrightarrow \mathcal{C}^{++}$ be a plus functor for $\mathcal{C}^+$. Then $+' \circ +$ is a globalization of $\mathcal{C}$.

<u>Remark 14.6</u>: Every mathematical theory must swallow its predecessor. Ethics demands that the classical examples be expressible in this language. Consider the case of the category of schemes.

Let $\mathbf{Ring}^0$ be the opposite category of rings, let $\mathbf{LR}$ be the category of locally ringed spaces, and let Φ be the functor which to each ring assigns its spectral sheaf. Let Sch be the classical category of schemes, which appears as a subcategory of $\mathbf{LR}$. The category $\mathbf{LR}$ has a natural Grothendieck topology, and this topology meets the axioms of a global structure. In a natural manner, the topology pulls back along Φ to a topology on $\mathbf{Ring}^0$; this topology meets the axioms of a local structure. The functor Φ becomes a CLCS functorial embedding. Both topologies meet the doubling hypothesis.

Suppose $\varepsilon : \mathbf{Ring}^0 \longrightarrow \mathcal{E}$ is a choice of double plus functor ($\mathcal{C} \longrightarrow \mathcal{C}^{++}$). The universal properties imply existence (and, up to functorial equivalence, uniqueness) of a continuous functor $\Psi : \mathcal{E} \longrightarrow \mathbf{LR}$ such that $\Psi \circ \varepsilon$ is equivalent to Φ. Let $\mathcal{O}$ be the class of objects $A \in \mathbf{LR}$ for which there is $B \in \mathcal{E}$ and a $\mathbf{LR}$-isomorphism $A \longrightarrow \Psi(B)$. Let $\mathcal{S}$ be the full subcategory of **LR** whose object class is $\mathcal{O}$. By inspection, $\mathcal{S}$ with the subset topology of **LR** is a global structure on $\mathcal{S}$. Because Φ is a CLCS functorial embedding into $\mathcal{S}$, we conclude that $\Psi : \mathcal{E} \longrightarrow \mathcal{S}$, is a CLCS functorial embedding. Thus, Ψ is equivalent to ε! To show that the usual **Sch** meets our conditions for a globalization, it suffices to show that its object class is $\mathcal{O}$.

Every member of $\mathcal{O}$ has an affine cover, and so must be in **Sch**. The two-step nature of the abstract construction reveals itself in the proof of the reverse inequality. It is given that $\mathcal{O}$ is closed under descent; we just apply this fact twice. Suppose $S \in$ **Sch** and $\mathcal{U}$ is an affine cover of its base space. If $S_{|U \cap V}$ were in $\mathcal{O}$ for each $U,V \in \mathcal{U}$, then we could conclude that $S \in \mathcal{O}$. Thus, we are reduced to proving that schemes which are restrictions of affine elements to open subsets are in $\mathcal{O}$. Now suppose S happens to be a restriction of an affine. Elementary properties of the spectral topology assure that there exists an affine cover $\mathcal{U}$ of S such that $S_{|U \cap V}$ is actually affine for each $U,V \in \mathcal{U}$.

Remark 14.7: Many classical theories support a notion of coherent sheaf of modules. A theory of modules would, at first glance, require a notion of base space. In fact, the universal framework easily embraces such ideas.

Again, let us use schemes to illustrate. Let $\mathbf{Ring}^0$ and **Sch** be as before. (When studying coherent modules, one would probably restrict $\mathbf{Ring}^0$ to the category of finitely generated algebras over a fixed Noetherian ring.) Let **Mod** be the category in which

(14.8.a) a **Mod**-object is a pair (A,M) where $A \in \mathbf{Ring}^0$ and M is an A-module, and

(14.8.b) a **Mod**-morphism $(A,M) \longrightarrow (B,N)$ is pair $(\beta : B \longrightarrow A, f : N \longrightarrow M)$ where β is a ring homomorphism, f is an additive homomorphism, and $f(r \cdot x) = \beta(r) \cdot f(x)$ for $r \in B$ and $x \in N$,

(14.8.c) composition is defined using opposite conventions.

Define a universe of formal subsets for **Mod** to consist of all pairs (β,f) where β is a formal subset of $\mathbf{Ring}^0$ and f is tensor product with β. In a canonical way, the topology of $\mathbf{Ring}^0$ can be lifted to **Mod** so that **Mod** becomes a local structure and the forgetful functor $(A,M) \longrightarrow A$ becomes a continuous functor $\pi : \mathbf{Mod} \longrightarrow \mathbf{Ring}^0$. Let **Coh** be a globalization of **Mod**, and let π^* denote the canonical extension of π.

The category **Coh** is the category of coherent sheaves of modules. The unversal theory applies; for example, it can be used to show that the construction of dual modules lifts to **Coh**$\longrightarrow$**Coh**. If desireable, π can be regarded as a fiber functor which divides **Mod** into subcategoires; universal properties concerning the division π (such as the constraint that each fiber be an abelian category) can be lifted to the fibering induced from π^*, by an argument which applies to other examples as well.

Remark 14.9: Obviously, Remark 14.6 adapts to many classical cases. However, one must be careful about Hausdorff, or separation, conditions.

Let **EOpen** be the category of open subsets of Euclidean space, and let **Man** be the category of manifolds. Consider the following canopy $\mathfrak{G}$, of type Int({1,2}) inside **EOpen**:

(14.10) $\mathfrak{G}(1) = \mathfrak{G}(2) = \mathfrak{G}(1,1) = \mathfrak{G}(2,2) = \mathbb{R}$,

$\mathfrak{G}(1,2) = \mathfrak{G}(2,1) = \mathbb{R} - \{0\}$,

all $\mathfrak{G}$-morphisms are subset injection.

Although $\mathbb{R}$ is a colimit of $\mathfrak{G}$ in **EOpen**—for that matter, in **Man** as well—the corresponding cone fails to be an affinization. The true affinization of $\mathfrak{G}$ is the famous 'line with two origins', a non-Hausdorff, yet locally Euclidean, object. The category of all locally Euclidean spaces is the globalization of **EOpen**; **Man** is merely a subcategory. But, **Man** is the category of objects of true interest.

We do not claim (nor believe) that there is a universal method for selecting 'Hausdorff' objects from an arbitrary global category. However, given a category $\mathfrak{C}$ with some notion of Hausdorff, there is a simple way to lift it to $\mathfrak{C}^+$. Although we do not pursue the theme here, general propositions about Hausdorffness—such as closure under finite inverse limits—can be proved.

Actually, algebraic geometry has already revised the Hausdorff property. After all, the correct generalization of the classical algebraic variety is not the scheme, but rather the *separated* scheme. Recall that a morphism $b : B \longrightarrow A$ is separated provided that the diagonal $B \longrightarrow (B,b) \times_A (B,b)$ is a *closed embedding*. Thus, definition of separation inside an arbitrary category reduces to choosing a suitable notion of closed embedding.

Let $\mathfrak{C}$ be a local structure. Suppose a class CEmb of closed embeddings for $\mathfrak{C}$ has been selected. At present, and in the author's opinion, this choice depends very much on the specific $\mathfrak{C}$ involved. Formally, all we require of CEmb is that it be a universe of layered morphisms with respect to the topology. Now let $+ : \mathfrak{C} \longrightarrow \mathfrak{C}^+$ be the plus functor. Call a $\mathfrak{C}^+$-morphism $b : B \longrightarrow A$ a closed embedding *in* $\mathfrak{C}^+$ if there is an affine cover S of A such that, for each $s \in S$, $s^{-1}b$ can be represented by f^+ where $f \in$ CEmb. Because CEmb is a universe of layered morphisms, for $f \in Mor(\mathfrak{C})$, f is a closed embedding in $\mathfrak{C}$ if and only if f^+ is a closed embedding with respect to $\mathfrak{C}^+$. Hence, even if CEmb depends in a non-trivial way on $\mathfrak{C}$, it has a *canonical* extension to a notion of closed embedding for $\mathfrak{C}^+$.

Consider the case in which $\mathfrak{C}$ is **EOpen**, and CEmb is the class of closed immersions. The subset injection of unit circle into $\mathbb{R}^2$ is a closed embedding in the category of locally Euclidean spaces. This means that the circle is an affinization of a pullback system into the canopy for some cover of $\mathbb{R}^2$. Thus, although CEmb is a universe of layered morphisms, **EOpen** is not closed under it. Passing from $\mathfrak{C}$ to $\mathfrak{C}^+$, we get new closed embeddings into affine objects. This increase is fine for CEmb, but it must be avoided when lifting the class Cvm of covering morphisms.

Remark 14.11: Our guiding intuition has been that formal subsets generalize open embeddings. However, there are theories in which cut-and-paste is practiced with closed embeddings. As an example suggests, there is no formal distinction.

Let **Sim** be the category of simplices of all dimensions. The category **CW** of CW-complexes is a globalization of **Sim**, where a complex represents a family of simplices glued along closed subspaces. Clearly, the class CSub of closed embeddings in **Sim** is a universe of formal subsets. The usual notion of cover translates to a topology on CSub, and **Sim** becomes a local structure. However, this is not yet enough.

Let **Top** denote the category of topological spaces, and let $\Phi : \mathbf{Sim} \longrightarrow \mathbf{Top}$ be the canonical functor which sends a simplex to the corresponding cell. We would like to cite the universal Theorem 14.4 to show that Φ has a canonical extension $\mathbf{CW} \longrightarrow \mathbf{Top}$. However, the usual topology on **Top** is derived from open embeddings, and Φ fails to be continuous with respect to it. The solution is to observe that there is another global structure (Sub',Cov') on **Top** in which Sub' is the class of closed embeddings, and with respect to which Φ is continuous. The definition of Cov' is not quite obvious. If S is a cone of closed embeddings into a space X, one traditionally regards S as a cover if

(14.12) $\quad \cup_{s \in S} Im(s) = X.$

But, there are infinite cones S which satisfy (14.12) and which are not intrinsic. Instead, we say a cone $S \subseteq$ Sub' into $X \in$ **Top** belongs to Cov' if and only if there is a finite, non-empty $S' \subseteq S$ such that $\cup_{s \in S'} Im(s) = X$.

In **Top**, the topologies of open embeddings and of closed embeddings appear to be two aspects of one structure. The author regards this duality as an accident unique to **Top**. There does not appear to be a universal way to build a complement to a given categorical topology. On the other hand, the trick of defining a cover, in the formal sense, of closed embeddings to be a cone which contains a finite subcover, in the traditional sense, applies to other examples. The essential difference between topologies of open embeddings and those of closed embeddings seems to be that covers by the latter are basicly finite.

The present use of closed embeddings is wildly different from the interpretation of Remark 14.9. Gluing along closed embeddings is the same, formally, as gluing along open embeddings; each type of morphism is an acceptable kind of formal subset. To define separation, one needed a topology with formal subsets *and a completely distinct* notion of closed embedding to supplement it.

§15 Lifting Layered Morphisms

Throughout this section, assume

(15.1) $\mathfrak{C}$ is a local structure,

Lay is a universe of layered morphisms for $\mathfrak{C}$,

a choice of plus functor is fixed for $\mathfrak{C}$, and

$\mathfrak{C}$ is closed under Lay.

The closing lemmas of Section 12 show that a $\mathfrak{C}^p$-pseudoisomorphism is an absolute reduction. Also, recall that for $f,g \in \mathfrak{C}^p$, $f^s = g^s$ implies $f = g$.

Corollary 15.2: Let $A \in \mathfrak{C}$, $B_0 \in \mathfrak{C}^p$ and $\alpha \in \mathrm{Mor}_{\mathfrak{C}^+}(B_0, A^p)$. Then there is $f \in \mathrm{Mor}_{\mathfrak{C}^p}(B_0, A^p)$ so that $\alpha = f^s$.

Proof: There is $C_0 \in \mathfrak{C}^p$, $\Theta_0 \in \mathrm{Mor}_{\mathfrak{C}^p}(C_0, B_0)$ and $f_0 \in \mathrm{Mor}_{\mathfrak{C}^p}(C_0, A^p)$ so

(15.3.a) $f_0{}^s = \alpha \circ \Theta_0{}^s$,

(15.3.b) Θ_0 is a pseudoisomorphism.

As A^p is affine and Θ_0 is an absolute reduction, there is $f \in \mathrm{Mor}_{\mathfrak{C}^p}(B_0, A^p)$ such that $f_0 = f \circ \Theta_0$. Then $f^s = \alpha$. □

Definition 15.4: A $\mathfrak{C}^+$-morphism $b_1 : B_0 \longrightarrow A_0$ is called locally Lay if for each $j \in \Lambda(A_0)$ there is $c \in \mathrm{Lay}$ so $\mathrm{cod}\, c = A_0(j)$ and c^+ is a pullback of b_1 along the j-th coordinate chart into A_0. A $\mathfrak{C}^p$-morphism $b_0 : B_0 \longrightarrow C_0$ is said to be pseudo-Lay if $b_0{}^s$ is locally Lay. For the remainder of the section, denote the class of locally Lay morphisms by Lay^+.

Previous lemmas immediate provide some structure theory for locally Lay and pseudo-Lay morphisms.

(15.5.a) Let $b_1 \in \mathrm{Lay}^+$ and let α be a $\mathfrak{C}^+$-isomorphism such that $\mathrm{cod}\, \alpha = \mathrm{dom}\, b$. Then $b \circ \alpha \in \mathrm{Lay}^+$.

(15.5.b) Suppose $A,B \in \mathfrak{C}$, $b \in \mathrm{Lay}(B,A)$, $C_0 \in \mathfrak{C}^p$ and $c_1 \in \mathrm{Mor}_{\mathfrak{C}^+}(C_0{}^s, B^+)$ is a $\mathfrak{C}^+$-isomorphism. There is $c_0 \in \mathrm{Mor}_{\mathfrak{C}^+}(C_0, B^p)$ so $c_1 = c_0{}^s$. Thus, the locally Lay morphism $b^+ \circ c_1$ is $f_0{}^s$ where f_0 is a composition of an affinization with a Lay-morphism. Conversely, such a composition is pseudo-Lay.

Consequently,

(15.6.a) a pseudo-Lay morphism is a $\mathcal{C}^p$-pullback base,

(15.6.b) each pullback of a pseudo-Lay morphism along an affine morphism is pseudo-Lay.

(15.6.c) a locally Lay morphism is a $\mathcal{C}^+$-pullback base.

To proceed, the assumption that $\mathcal{C}$ is closed under Lay must be exploited.

Suppose $A \in \mathcal{C}$ and Θ is a cover of A. Let $J = \mathrm{dom}(\Theta)$, and let $\Theta^{\#} : A_0 \longrightarrow A$ be a canopy of Θ. Let $A_1 \in (\mathcal{C}^p)^p$ denote the image of A_0, regarded as a graph, under the pasting functor. Suppose (Q_1,q_1) is a pullback system of A_1 such that each value of q_1 is pseudo-Lay. For each $\beta \in J \amalg J^2$, fix an affinization $\beta^{\#} : Q_0(\beta) \longrightarrow \beta_0 \in \mathcal{C}$. Let Q_0 denote the graph of $\mathcal{C}$-objects of type Int(J) where

(15.7) $Q_0(\beta) = \beta_0$ for $\beta \in J \amalg J^2$ and

$Q_0(\beta,\rho,k)^p \circ \beta^{\#} = k^{\#} \circ Q_1(\beta,\rho,k)$ for $(\beta,\rho,k) \in \mathrm{Mor}(\mathrm{Int}(J))$.

For $j \in J$, let $q_0(j)$ be the affinization of $q_1(j)$ through $\beta^{\#}$. By hypothesis, $q_0(j) \in \mathrm{Lay}$ for each $j \in J$. Let Q' denote the image of Q_0 under pasting. Let $\varphi : Q_1 \longrightarrow B_0$ be a colimit in $\mathcal{C}^p$.

Part III characterizes the interplay between affinization and inverse limits. From Lemma 2.46 and Corollary 8.18, it follows that Q_0 is a canopy over $\mathcal{C}$. Now $\tau : j \mapsto [1_{Q_0(j)},1]$ is a colimit $Q' \longrightarrow Q_0$, and there is a unique $b_0 \in \mathrm{Mor}_{\mathcal{C}^p}(B_0,Q_0)$ such that $b_0 \circ \varphi(j) = \tau(j) \circ j^{\#}$ for all $j \in J$. For $\beta \in J \amalg J^2$, $(\beta^{\#})^s$ is a $\mathcal{C}^+$-isomorphism. Consequently, b_0 is a pseudoisomorphism.

Again, Part III implies that (Q_0,q_0) is a pullback system of Lay-morphisms. Let q' be the $\mathcal{C}^p$-morphism $j \mapsto [q_0(j),j]$ on $j \in J$. Hypothesis (15.1) states that $\Theta^{\#} \circ q'$ is pseudo-Lay. If the original (Q_1,q_1) is derived from some morphism $c_0 : C_0 \longrightarrow A^p$, then (C_0,c_0) is $\mathcal{C}^p/A^p$-isomorphic to $\Theta^{\#} \circ q' \circ b_0$, which is also pseudo-Lay.

<u>Theorem 15.8:</u> Lay^+ is a universe of layered morphisms for $\mathcal{C}^+$ with the e.l.-topology. Moreover,

(15.9.a) for $A \in \mathcal{C}$, any locally Lay morphism into A^+ is $\mathcal{C}^+/A^+$-isomorphic to b+ where $b \in \mathrm{Lay}$,

(15.9.b) if Lay is a universe of embeddings for $\mathcal{C}$, then Lay^+ is a universe of embeddings for $\mathcal{C}^+$

(15.9.c) if Lay meets the smallness condition, then so does Lay^+.

Proof: For $A_0 \in \mathcal{C}^p$ and $b_0 \in \mathcal{C}^p/A_0$, b_0 is determined up to $\mathcal{C}^p/A_0$-morphism by its pullbacks of the coordinate charts of A_0; moreover, a list of pullbacks is achieved by a morphism if and only if they form a canopy. For b_0 to be pseudo-Lay, each pullback must be a composition of an affinization with a Lay-morphism. Two affinizations into a $\mathcal{C}$-object are isomorphic if and only if their corresponding covers mutually refine one another. Once the rest of the therem is established, condition (15.9.c) follows from these remarks.

Suppose $b_1 : B_1 \longrightarrow A_1$ is a locally Lay $\mathcal{C}^+$-morphism. Let $C \in \mathcal{C}$ and $c_1 \in \mathrm{Mor}_{\mathcal{C}^+}(C^+, A_1)$. We claim that $c_1^{-1}b_1$ is locally Lay. Without loss of generality, we may assume $b_1 = b_0^S$ for $b_0 \in \mathrm{Mor}(\mathcal{C}^p)$. There exist $C_0 \in \mathcal{C}^p$, $\Theta_0 \in \mathrm{Mor}_{\mathcal{C}^p}(C_0, C^p)$ and $c_0 \in \mathrm{Mor}_{\mathcal{C}^p}(C_0, A_1)$ such that Θ_0 is an affinization and $c_0^S = c_1 \circ \Theta_0^S$. Now $c_0^{-1}b_0$ is pseudo-Lay. The comments preceeding the theorem's statement apply to the pullback system of the assigned cover along $c_0^{-1}b_0$.

We may conclude that a morphism b_1 is in Lay^+ if and only if for each $C \in \mathcal{C}$ and each $c_1 \in \mathrm{Mor}_{\mathcal{C}^+}(C^+, \mathrm{cod}\, b_1)$ the pullback of b_1 along c_1 is $\mathcal{C}^+/C^+$-isomorphic to a Lay morphism under the plus functor. With this new characterization, it is clear that Lay^+ forms a universe of subsets. If Lay is a universe of embeddings, obviously Lay^+ must be as well.

Now suppose $b_0 : B_0 \longrightarrow A_0$ is a $\mathcal{C}^p$-morphism and Θ is a $\mathcal{C}^p$-cover of A_0 such that $\Theta(j)^{-1}b_0$ is pseudo-Lay for each $j \in \mathrm{dom}(\Theta)$. The remaining claims easily reduce to the statement that b_0 must be pseudo-Lay. It suffices to consider pullbacks of b_0 along coordinate charts; by pulling back Θ, we reduce to the case where $A_0 = A^p$ for $A \in \mathcal{C}$. Again, the argument before the theorem's statement assures that b_0 must be pseudo-Lay. □

Corollary 15.10: Let Iso be the set of $\mathcal{C}$-isomorphisms. Then Iso is a universe of layered morphisms for $\mathcal{C}$. Moreover, if Lay = Iso, then Lay^+ is the class of $\mathcal{C}^+$-isomorphisms.

Proof: Trivial. □

§16 The Plus Topology

We are ready to prove Theorem 14.4. Assume ($\mathfrak{C}$,Sub,Cov) is a local structure. Let $\mathrm{Cvm} = \mathrm{Cvm}_{\mathfrak{C}}$, let Cvm^+ be the class of all locally Cvm morphisms, and let $(\mathrm{Sub}^+,\mathrm{Cov}^+)$ be the cover reduction of the e.l.-topology through Cvm^+; throughout this section, regard it as the topology of $\mathfrak{C}^+$.

It is known that

(16.1.a) Cov^+ is flush, intrinsic, and meets the CLCS condition,

(16.1.b) for $A,B \in \mathfrak{C}$, the plus functor induces a bijection $\mathrm{Mor}_{\mathfrak{C}}(B,A) \longrightarrow \mathrm{Mor}_{\mathfrak{C}^+}(B^+,A^+)$,

(16.1.c) if Cvm is the class of $\mathfrak{C}$-isomorphisms, then Cvm^+ is the class of $\mathfrak{C}^+$-isomorphisms.

Suppose $b_0 : B_0 \longrightarrow A_0$ is a covering morphism in $\mathfrak{C}^p$. Each pullback of b_0 along a chart can be identified with a covering morphism of $\mathfrak{C}$. Consequently, the smoothing and plus functors are continuous. The CLCS condition for $\mathfrak{C}$ easily implies that for $b \in \mathrm{Mor}(\mathfrak{C})$,

(16.2) $b \in \mathrm{Sub} \iff b^+ \in \mathrm{Sub}^+$.

The assumption that $\mathfrak{C}$ is flush implies that for S a nonempty cone in $\mathfrak{C}$,

(16.3) $S \in \mathrm{Cov} \iff + \circ 1_S$ is an indexed $\mathfrak{C}^+$-cover.

Hence, the plus functor is a CLCS functorial embedding.

To proceed, we need a less abstract description of Sub^+-morphism. Fix $A_0 \in \mathfrak{C}^p$; the immediate problem is to classify Sub^+-morphisms into A_0 up to $\mathfrak{C}/A_0$-isomorphism. First,any formal subset into A_0 can be represented by a $\mathfrak{C}^p$-morphism $b_0 : B_0 \longrightarrow A_0$ such that $b_0(j) \in \mathrm{Sub}^p$ for each $j \in \Lambda\,(B_0)$. Put

$$\beta = \{\, b_0(j) : j \in \Lambda\,(B_0)\} \,.$$

Suppose Θ is a cone of formal $\mathfrak{C}^p$-subsets into A_0 such that β factos through Θ and Θ factors through β. (In particular, if K is a choice of representatives of Sub^p into A_0, we may choose Θ to be the set of members of K which are isomorphic to members of β.) The canopy of Θ consists of affine objects, and has an affinization C_0 in $\mathfrak{C}^p$. Let c_0 denote the canonical morphism $C_0 \longrightarrow A_0$. Now fix a fibered product $(P;\pi_B,\pi_C)$ for $(B_0,b_0)\times_{A_0}(C_0,c_0)$. The following observations are elementary:

(16.4.a) c_0 is a monomorphism. Consequently, π_B is monomorphic as well.

(16.4.b) Let $k \in \Lambda(C_0)$. Now $\Theta(k) = c_0 \circ \iota_k$ factors through b_0, and the image of b_0 under smoothing is in Sub^+. It follows that the pullback of π_C along ι_k is taken to a Cvm^+-morphism by smoothing.

(16.4.c) For each $j \in \Lambda(B_0)$, $b_0 \circ \iota_j$ factors through Θ. Using (16.4.a), it follows that there is a unique $\mathcal{C}^p$-morphism $b':B_0 \longrightarrow C_0$ such that $c_0 \circ b' = b_0$.

Tautologically, $\pi_B \circ (1_{B_0} \wedge b') = 1_{B_0}$. Since π_B is monomorphic, this is sufficient to imply that π_B is an isomorphism. The theory of the last section implies that $\pi_C{}^s$ is in Cvm^+. Thus, $b_0{}^s$ is isomorphic to a composition of a Cvm^+-morphism and a map of the kind c_0.

The above classification has several consequences. First, it is now trivial (and left to the reader) to check that Cov^+ meets the smallness condition. In addition, it also offers a criterion of use in proof of Theorem 14.4.(E):

Lemma 16.5: Suppose that the canopy of any cone of formal $\mathcal{C}$-subsets has an affinization in $\mathcal{C}$. Then $\mathcal{C}^+$ is closed under affinization.

Proof: Suppose G is a canopy of $\mathcal{C}^+$-objects. Let $\Theta_1:G_1 \longrightarrow G^p$ be a canopy derived from a $(\mathcal{C}^+)^p$-cover of G_0. Now G admits an affinization if and only if the colimit of G_1 (in $(\mathcal{C}^+)^p$) admits an affinization. Consequently, $\mathcal{C}^+$ will be closed under affinization if every canopy G, for which G(j) is affine (meaning here, in the image of +) for $j \in \Lambda(G)$, admits an affinization.

Tautologically, any canopy of $\mathcal{C}^+$ in which *every* member is affine has an affinization—namely, itself regarded as a member of $\mathcal{C}^p$ under smoothing! Hence, the lemma's conclusion will follow if we can show that for $A \in \mathcal{C}$, every formal $\mathcal{C}^+$-subset into A^+ is isomorphic to an affine subset. Now fix $A \in \mathcal{C}$, and represent a formal subset into A^+ as $c_0 \circ \pi_C$, as previously constructed. The hypothesis of the lemma states that C_0 is pseudoaffine. The theory of lifting finishes the proof by assuring that dom π_C must also be pseudoaffine. □

The next step in proof of Part E is to show that, for an arbitrary local structure $\mathcal{C}$, the category $\mathcal{D} = \mathcal{C}^+$ does satisfy the hypothesis of Lemma 16.5. The demonstration of this fact is an easy variation on the preceeding proof, and is left to the reader.

The last non-trivial obstruction to the Main Theorem is the claim that $\mathcal{C}^+$ must be closed under Cvm^+. Consider $A_1 \in \mathcal{C}^+$, $\Theta^{\#}:A_2 \longrightarrow A_1{}^p$ a canopy of a cover of A_1, and (Q_2,q_2)

a pullback system of Cvm^+-morphisms into A_2. Let q denote the $(\mathcal{C}^+)^p$-morphism $j \mapsto [q(j),j]$ on Q_0. We claim that $\Theta^{\#} \circ q$ is the composition of a Cvm^+-morphism with an affinization (in $(\mathcal{C}^+)^p$). Fix a choice of smoothing functor for $(\mathcal{C}^+)^p$.

If Q_0 admits an affinization, then the system is a pullback along a morphism from that affinization; it follows that $\Theta^{\#} \circ q$ is a composition of the desired type. Thus, it suffices to show that Q_0 is pseudoaffine. The established properties of $(\mathrm{Sub}^+,\mathrm{Cov}^+)$ allow us to replace Q_0 by any pullback of a pseudoisomorphism along q.

There exists a $(\mathcal{C}^+)^p$-cover φ of A_1 such that, for each $j \in \mathrm{dom}(\varphi)$, there is $A(j) \in \mathcal{C}$ and $\tau(j)$ a formal $\mathcal{C}^p$-subset so $\varphi(j) = (A(j)^+,\tau(j)^s)$. Put $J = \mathrm{dom}(\varphi)$. Let $\varphi^{\#}: A_2 \longrightarrow A_1$ be a canopy of φ in $\mathcal{C}^p$, where we choose every member of A_2 to be ($\mathcal{C}$-)affine. We may regard $\varphi^{\#}$ as a canopy over $\mathcal{C}^+$. There is a function $\beta \mapsto (B(\beta),g(\beta))$ on $\beta \in J \amalg J^2$ such that for each β,

(16.6) $\quad B(\beta) \in \mathcal{C}$ and $g(\beta) \in \mathrm{Cvm}(B(\beta),A_2(\beta))$,

and such that for $j,k \in J$,

(16.7.a) $\quad (\{B(j)^+\}^p,\{g(j)^+\}^p)$ is a pullback $\varphi(j)^{-1}q$,

(16.7.b) $\quad (\{B(j,k)^+\}^p,\{g(j,k)^+\}^p)$ is a pullback $\{A_2((j,k),\rho,j) \circ \varphi(j)\}^{-1}q$.

Using these pullbacks, we may build a graph Q' of $\mathcal{C}$-objects so that $(Q'^+, q' : j \mapsto g(j)^+)$ is a pullback system of $\varphi^{\#}$ along q. It suffices to show that Q' is pseudoaffine. But Q' is a canopy of $\mathcal{C}$, which means $(Q')^s$ serves as a colimit of $(Q')^+$.

The topology $(\mathrm{Sub}^+,\mathrm{Cov}^+)$ is now known to be a local structure. Clearly, it is the plus topology of Theorem 14.4.

Most of the Theorem 14.4.(A-E) falls from the machinery of earlier sections. Suppose $\Gamma : \mathcal{C}^+ \longrightarrow \mathcal{D}$ is a continuous functor such that $\Gamma \circ +$ is a CLCS functorial embedding. The conclusion that Γ is also a CLCS functorial embedding does not follow directly, but requires a simple argument in the style of what has gone before. Verification is left to the reader.

One last observation completes Theorem 15.8. That arguemnt describes the lift of a universe of layered morphisms Lay under which $\mathcal{C}$ is closed. At the time, we only had the e.l.-topology for $\mathcal{C}^+$, and it is not necessarily true that this topology is closed under Lay^+. However, the plus topology has fewer formal subsets; this fact was used in our proof that $\mathcal{C}^+$ is closed under Cvm^+. An obvious modification of the latter argument yields:

<u>Corollary 16.8:</u> Let $\mathfrak{C}$ be a local structure. Let Lay be a universe of layered morphisms for $\mathfrak{C}$, under which that category is closed. Let Lay^{+} be the lift of Lay to $\mathfrak{C}^{+}$. Then, with respect to the plus topology, $\mathfrak{C}^{+}$ is closed under Lay^{+}.

References

[D] A. Douady, *Le Problème des Modules pour les Sous-Espaces Analytiques Compacts d'un espace Analytique donné*, Ann. Inst. Fourier, Grenoble **16** 1 (1966) p1-95

[Gö] K. Gödel, *The consistency of the continuum hypothesis*, Studies, Ann.. of Math., No. 3 Princeton: Princeton University, 1940

[Gr] A. Grothendieck, *Revetements étales et groupe fondamentale*, (SGA 1), Springer Lecture Notes in Math **224**, 1971

[H] R. Hartshorne, *Algebraic Geometry*, Springer-Verlag: New York, 1977

[J] P.T. Johnstone, *Topos Theory*, Academic Press: New York, 1977

[K] D. Knutson, *Algebraic Spaces*, Lect. Notes in Math. 203, Springer-Verlag: New York, 1971

[Ma] S. MacLane, *Categories for the Working Mathematician*, Springer-Verlag: New York, 1988

[Mi] J.S. Milne, *Étale Cohomology*, Princeton University Press: New York, 1977

[S] I.R. Shafarevich, *Basic Algebraic Geometry*, Springer-Verlag: New York, 1977

Paul Feit

University of Toledo

Toledo, Ohio

Editorial Information

To be published in the *Memoirs*, a paper must be correct, new, nontrivial, and significant. Further, it must be well written and of interest to a substantial number of mathematicians. Piecemeal results, such as an inconclusive step toward an unproved major theorem or a minor variation on a known result, are in general not acceptable for publication. *Transactions* Editors shall solicit and encourage publication of worthy papers. Papers appearing in *Memoirs* are generally longer than those appearing in *Transactions* with which it shares an editorial committee.

As of November 1, 1992, the backlog for this journal was approximately 9 volumes. This estimate is the result of dividing the number of manuscripts for this journal in the Providence office that have not yet gone to the printer on the above date by the average number of monographs per volume over the previous twelve months. (There are 6 volumes per year, each containing about 3 or 4 numbers.)

A Copyright Transfer Agreement is required before a paper will be published in this journal. By submitting a paper to this journal, authors certify that the manuscript has not been submitted to nor is it under consideration for publication by another journal, conference proceedings, or similar publication.

Information for Authors

Memoirs are printed by photo-offset from camera copy fully prepared by the author. This means that the finished book will look exactly like the copy submitted.

The paper must contain a *descriptive title* and an *abstract* that summarizes the article in language suitable for workers in the general field (algebra, analysis, etc.). The *descriptive title* should be short, but informative; useless or vague phrases such as "some remarks about" or "concerning" should be avoided. The *abstract* should be at least one complete sentence, and at most 300 words. Included with the footnotes to the paper, there should be the 1991 *Mathematics Subject Classification* representing the primary and secondary subjects of the article. This may be followed by a list of *key words and phrases* describing the subject matter of the article and taken from it. A list of the numbers may be found in the annual index of *Mathematical Reviews*, published with the December issue starting in 1990, as well as from the electronic service e-MATH [**telnet e-MATH.ams.org** (or **telnet 130.44.1.100**). Login and password are **e-math**]. For journal abbreviations used in bibliographies, see the list of serials in the latest *Mathematical Reviews* annual index. When the manuscript is submitted, authors should supply the editor with electronic addresses if available. These will be printed after the postal address at the end of each article.

Electronically-prepared manuscripts. The AMS encourages submission of electronically-prepared manuscripts in AMS-TEX or AMS-LATEX. To this end, the Society has prepared "preprint" style files, specifically the amsppt style of AMS-TEX and the amsart style of AMS-LATEX, which will simplify the work of authors and of the production staff. Those authors who make use of these style files from the beginning of the writing process will further reduce their own effort.

Guidelines for Preparing Electronic Manuscripts provide additional assistance and are available for use with either $\mathcal{A}_\mathcal{M}\mathcal{S}$-TEX or $\mathcal{A}_\mathcal{M}\mathcal{S}$-LATEX. Authors with FTP access may obtain these *Guidelines* from the Society's Internet node e-MATH.ams.org (130.44.1.100). For those without FTP access they can be obtained free of charge from the e-mail address guide-elec@math.ams.org (Internet) or from the Publications Department, P. O. Box 6248, Providence, RI 02940-6248. When requesting *Guidelines* please specify which version you want.

Electronic manuscripts should be sent to the Providence office only after the paper has been accepted for publication. Please send electronically prepared manuscript files via e-mail to pub-submit@math.ams.org (Internet) or on diskettes to the Publications Department address listed above. When submitting electronic manuscripts please be sure to include a message indicating in which publication the paper has been accepted.

For papers not prepared electronically, model paper may be obtained free of charge from the Editorial Department at the address below.

Two copies of the paper should be sent directly to the appropriate Editor and the author should keep one copy. At that time authors should indicate if the paper has been prepared using $\mathcal{A}_\mathcal{M}\mathcal{S}$-TEX or $\mathcal{A}_\mathcal{M}\mathcal{S}$-LATEX. The *Guide for Authors of Memoirs* gives detailed information on preparing papers for *Memoirs* and may be obtained free of charge from AMS, Editorial Department, P.O. Box 6248, Providence, RI 02940-6248. The *Manual for Authors of Mathematical Papers* should be consulted for symbols and style conventions. The *Manual* may be obtained free of charge from the e-mail address cust-serv@math.ams.org or from the Customer Services Department, at the address above.

Any inquiries concerning a paper that has been accepted for publication should be sent directly to the Editorial Department, American Mathematical Society, P. O. Box 6248, Providence, RI 02940-6248.

Editors

This journal is designed particularly for long research papers (and groups of cognate papers) in pure and applied mathematics. Papers intended for publication in the *Memoirs* should be addressed to one of the following editors:

Recent Titles in This Series

(Continued from the front of this publication)

453 **Alfred S. Cavaretta, Wolfgang Dahmen, and Charles A. Micchelli,** Stationary subdivision, 1991

452 **Brian S. Thomson,** Derivates of interval functions, 1991

451 **Rolf Schön,** Effective algebraic topology, 1991

450 **Ernst Dieterich,** Solution of a non-domestic tame classification problem from integral representation theory of finite groups ($\Lambda = RC_3, v(3) = 4$), 1991

449 **Michael Slack,** A classification theorem for homotopy commutative H-spaces with finitely generated mod 2 cohomology rings, 1991

448 **Norman Levenberg and Hiroshi Yamaguchi,** The metric induced by the Robin function, 1991

447 **Joseph Zaks,** No nine neighborly tetrahedra exist, 1991

446 **Gary R. Lawlor,** A sufficient criterion for a cone to be area-minimizing, 1991

445 **S. Argyros, M. Lambrou, and W. E. Longstaff,** Atomic Boolean subspace lattices and applications to the theory of bases, 1991

444 **Haruo Tsukada,** String path integral realization of vertex operator algebras, 1991

443 **D. J. Benson and F. R. Cohen,** Mapping class groups of low genus and their cohomology, 1991

442 **Rodolfo H. Torres,** Boundedness results for operators with singular kernels on distribution spaces, 1991

441 **Gary M. Seitz,** Maximal subgroups of exceptional algebraic groups, 1991

440 **Bjorn Jawerth and Mario Milman,** Extrapolation theory with applications, 1991

439 **Brian Parshall and Jian-pan Wang,** Quantum linear groups, 1991

438 **Angelo Felice Lopez,** Noether-Lefschetz theory and the Picard group of projective surfaces, 1991

437 **Dennis A. Hejhal,** Regular b-groups, degenerating Riemann surfaces, and spectral theory, 1990

436 **J. E. Marsden, R. Montgomery, and T. Ratiu,** Reduction, symmetry, and phase mechanics, 1990

435 **Aloys Krieg,** Hecke algebras, 1990

434 **François Treves,** Homotopy formulas in the tangential Cauchy-Riemann complex, 1990

433 **Boris Youssin,** Newton polyhedra without coordinates Newton polyhedra of ideals, 1990

432 **M. W. Liebeck, C. E. Praeger, and J. Saxl,** The maximal factorizations of the finite simple groups and their automorphism groups, 1990

431 **Thomas G. Goodwillie,** A multiple disjunction lemma for smooth concordance embeddings, 1990

430 **G. M. Benkart, D. J. Britten, and F. W. Lemire,** Stability in modules for classical Lie algebras: A constructive approach, 1990

429 **Etsuko Bannai,** Positive definite unimodular lattices with trivial automorphism groups, 1990

428 **Loren N. Argabright and Jesús Gil de Lamadrid,** Almost periodic measures, 1990

427 **Tim D. Cochran,** Derivatives of links: Milnor's concordance invariants and Massey's products, 1990

426 **Jan Mycielski, Pavel Pudlák, and Alan S. Stern,** A lattice of chapters of mathematics: Interpretations between theorems, 1990

425 **Wiesław Pawłucki,** Points de Nash des ensembles sous-analytiques, 1990

424 **Yasuo Watatani,** Index for C^*-subalgebras, 1990

(See the AMS catalog for earlier titles)